S0-BBN-515

Experimenting with Everyday Science

Tools and Machines

Experimenting with Everyday Science

Art and Architecture

Food

Man-made Materials

Music

Sports

Tools and Machines

Experimenting with Everyday Science

Tools and Machines

Stephen M. Tomecek

To my good friend, Bill Boyle—a man who really knows his machines.
Thanks for being such a great teacher!

Chelsea House
An imprint of Infobase Publishing
132 West 31st Street
New York NY 10001

Library of Congress Cataloging-in-Publication Data
Tomecek, Steve.
Tools and machines / by Stephen M. Tomecek.
p. cm. — (Experimenting with everyday science)
Includes bibliographical references and index.
ISBN 978-1-60413-171-0 (hardcover)
1. Tools—Juvenile literature. 2. Machinery—Juvenile literature. I. Title. II. Series.

TJ1195.T615 2010
621.9—dc22 2009022332

Chelsea House books are available at special discounts when purchased in bulk quantities for businesses, associations, institutions, or sales promotions. Please call our Special Sales Department in New York at (212) 967-8800 or (800) 322-8755.

You can find Chelsea House on the World Wide Web at http://www.chelseahouse.com

Text design and composition by Annie O'Donnell
Cover design by Alicia Post
Cover printed by Bang Printing, Brainerd, MN
Book printed and bound by Bang Printing, Brainerd, MN
Date printed: April 2010
Printed in the United States of America

10 9 8 7 6 5 4 3 2 1

This book is printed on acid-free paper.

Contents

Introduction

When you hear the word *science*, what's the first thing that comes to mind? If you are like most people, it's probably an image of a laboratory filled with tons of glassware and lots of sophisticated equipment. The person doing the science is almost always wearing a white lab coat and is probably looking rather serious while engaged in some type of experiment. While there are many places where this traditional view of a scientist still holds true, labs aren't the only place where science is at work. Science can also be found at a construction site, on a basketball court, and at a concert by your favorite band. The truth of the matter is that science is happening all around us. It's at work in the kitchen when we cook a meal, and we can even use it when we paint a picture. Architects use science when they design a building, and science also explains why your favorite baseball player can hit a home run.

In **Experimenting with Everyday Science**, we are going to examine some of the science that we use in our day-to-day lives. Instead of just talking about the science, these books are designed to put the science right in your hands. Each book contains about 25 experiments centering on one specific theme. Most of the materials used in the experiments are things that you can commonly find around your house or school.

Once you are finished experimenting, it is our hope that you will have a better understanding of how the world around you works. While reading these books may not make you a world-class athlete or the next top chef, we hope that they inspire you to discover more about the science behind everyday things and encourage you to make the world a better place!

Safety Precautions

REVIEW BEFORE STARTING ANY EXPERIMENT

Each experiment includes special safety precautions that are relevant to that particular project. These do not include all the basic safety precautions that are necessary whenever you are working on a scientific experiment. For this reason, it is necessary that you read and remain mindful of the General Safety Precautions that follow.

Experimental science can be dangerous, and good laboratory procedure always includes carefully following basic safety rules. Things can happen very quickly while you are performing an experiment. Materials can spill, break, or even catch fire. There will be no time after the fact to protect yourself. Always prepare for unexpected dangers by following the basic safety guidelines during the entire experiment, whether or not something seems dangerous to you at a given moment.

We have been quite sparing in prescribing safety precautions for the individual experiments. For one reason, we want you to take very seriously every safety precaution that is printed in this book. If you see it written here, you can be sure that it is here because it is absolutely critical.

Read the safety precautions here and at the beginning of each experiment before performing each activity. It is difficult to remember a long set of general rules. By rereading these general precautions every time you set up an experiment, you will be reminding yourself that lab safety is critically important. In addition, use your good judgment and pay close attention when performing potentially dangerous procedures. Just because the text does not say "be careful with hot liquids" or "don't cut yourself with a knife" does not mean that you can be careless when boiling water or punching holes in plastic bottles. Notes in the text are special precautions to which you must pay special attention.

GENERAL SAFETY PRECAUTIONS

Accidents caused by carelessness, haste, insufficient knowledge, or taking an unnecessary risk can be avoided by practicing safety procedures and being alert while conducting experiments. Be sure to check the individual experiments in this book for additional safety regulations and adult supervision requirements. If you will be working in a lab, do not work alone. When you are working off-site, keep in groups with a minimum of three students per group, and follow school rules and state legal requirements for the number of supervisors required. Ask an adult supervisor with basic training in first aid to carry a small first-aid kit. Make sure everyone knows where this person will be during the experiment.

PREPARING

- Clear all surfaces before beginning experiments.
- Read the instructions before you start.
- Know the hazards of the experiments and anticipate dangers.

PROTECTING YOURSELF

- Follow the directions step-by-step.
- Do only one experiment at a time.
- Locate exits, fire blanket and extinguisher, master gas and electricity shut-offs, eyewash, and first-aid kit.
- Make sure there is adequate ventilation.
- Do not horseplay.
- Keep floor and workspace neat, clean, and dry.
- Clean up spills immediately.
- If glassware breaks, do not clean it up; ask for teacher assistance.
- Tie back long hair.
- Never eat, drink, or smoke in the laboratory or workspace.
- Do not eat or drink any substances tested unless expressly permitted to do so by a knowledgeable adult.

USING EQUIPMENT WITH CARE

- Set up apparatus far from the edge of the desk.
- Use knives or other sharp-pointed instruments with care.
- Pull plugs, not cords, when removing electrical plugs.
- Clean glassware before and after use.
- Check glassware for scratches, cracks, and sharp edges.
- Clean up broken glassware immediately.
- Do not use reflected sunlight to illuminate your microscope.
- Do not touch metal conductors.
- Use alcohol-filled thermometers, not mercury-filled thermometers.

USING CHEMICALS

- Never taste or inhale chemicals.
- Label all bottles and apparatus containing chemicals.
- Read labels carefully.
- Avoid chemical contact with skin and eyes (wear safety glasses, lab apron, and gloves).
- Do not touch chemical solutions.
- Wash hands before and after using solutions.
- Wipe up spills thoroughly.

HEATING SUBSTANCES

- Wear safety glasses, apron, and gloves when boiling water.
- Keep your face away from test tubes and beakers.
- Use test tubes, beakers, and other glassware made of Pyrex glass.
- Never leave apparatus unattended.
- Use safety tongs and heat-resistant gloves.

- If your laboratory does not have heat-proof workbenches, put your Bunsen burner on a heat-proof mat before lighting it.
- Take care when lighting your Bunsen burner; light it with the airhole closed, and use a Bunsen burner lighter in preference to wooden matches.
- Turn off hot plates, Bunsen burners, and gas when you are done.
- Keep flammable substances away from flames and other sources of heat.
- Have a fire extinguisher on hand.

FINISHING UP

- Thoroughly clean your work area and any glassware used.
- Wash your hands.
- Be careful not to return chemicals or contaminated reagents to the wrong containers.
- Do not dispose of materials in the sink unless instructed to do so.
- Clean up all residues and put them in proper containers for disposal.
- Dispose of all chemicals according to all local, state, and federal laws.

BE SAFETY CONSCIOUS AT ALL TIMES!

The Need for Tools

Humans are truly amazing creatures. Compared with many other animals on the planet, it might seem as if we are at a serious disadvantage. Ostriches can outrun us, elephants can out-pull us, and lions can rip us to pieces. Bats have better hearing, eagles have better eyesight, and even mosquitoes and wasps pack more of a sting. Despite all these limitations, we've done pretty well as a species. That's because unlike most other animals, humans can make **tools**. Using tools, humans have molded and shaped the world. Tools also have allowed us to adapt to just about every environment on the planet. The truth of the matter is that without tools, humans probably would have become extinct a very long time ago.

When people hear the word *tool*, they usually think of objects such as hammers, screwdrivers, or even chain saws. Tools aren't just for building, though. In fact, most of us can't go more than a few minutes without using some type of tool. When you brush your teeth, you use a tool. When you eat your cereal or butter a bagel, you are using tools. Even pencils and pens are specialized tools. In the most general sense, a tool can be any device used to get a job done or make a task easier. Some tools, such as a sewing needle or a wrench, are simple. They require only the force of a human hand to make them work. Other tools, such as a table saw or a drill press, are complex. They have many moving parts, and motors or engines power them.

These days, people have an incredible variety of tools at their disposal to do all sorts of different jobs. If you take a trip to a local hardware store or home improvement store, you can find hundreds of different devices for doing all sorts of tasks. Some tools, such as a pair of pliers, are versatile: They can be used for many different jobs. Others, such as a torque wrench, are designed for only one specific task. Regardless of a tool's size, use, or complexity, our ability to make and use tools has been the driving force behind human evolution. Tools have taken us from being simple nomadic scavengers to space explorers and movers of mountains.

No one knows for certain how long humans have been making and using tools, but many scientists believe that we've been at it for at least 2.5 million years. We don't have an exact date because we're not sure what the first tools looked like. In addition, the historical record from this time is very spotty, so just finding true artifacts is very difficult.

Some of the earliest tools were made out of stone. They look similar to rocks that you might find lying around on the ground. In fact, unless you were trained in identifying tools, you would probably walk right over these rocks without even noticing them. Many scientists believe that before making tools, early humans began using stones for different tasks. Small rounded rocks were used to crack open nuts or break animal bones to get the marrow out. Rocks with sharp edges were used to cut through animal skins and tough plant material, such as vines.

Over time, humans discovered that they didn't have to waste time looking for stones with the right shapes. Instead, they found that they could use one stone to chip away at another stone and shape it to fit the job. They learned that they could make tools to help them in their everyday lives.

THE FIRST "HANDY MAN"

The ability to intentionally create a tool for a specific purpose seems to be one of those traits that set humans apart from the rest of the animal kingdom. One of the first human species that made tools is *Homo habilis*. *H. habilis* lived about 2 million years ago. In 1960, Mary and Louis Leakey discovered the first *H. habilis* fossils at Olduvai Gorge in Tanzania. The Leakeys had been working at Olduvai Gorge for several years. They had found many simple tools that had been chipped out of stone. When they finally found fossil bones to go with the tools, Louis was convinced that he had found the remains of one of the earliest toolmaking humans. At the suggestion of another scientist named Raymond Dart, he gave the species the name ***Homo***

Homo habilis was one of the earliest toolmakers. This species made simple tools out of stone. In fact, *H. habilis's* remains in Tanzania and Kenya are often found buried near primitive stone tools these primates crafted.

habilis, which in Latin means "handy man." Many scientists question whether *Homo habilis* was truly the first toolmaker or even a direct ancestor of modern humans. In any case, there is evidence that humans have been making tools for a very long time.

Based on the archaeological record, it appears that the first manufactured tools were simple choppers created from rounded, fist-sized stones. Making a chopper is a fairly easy process. Hitting one edge of a stone with another stone of equal or greater hardness creates a jagged edge that could be used for cutting animal hides, sinews, small tree branches, and vines. **Experiment 1: *The Wedge Design of a Stone Chopper*** shows how effective a stone chopper is at cutting natural materials.

EXPERIMENT 1 The Wedge Design of a Stone Chopper

Topic

Can a simple stone chopper be an effective cutting tool?

Introduction

These days, humans have many tools for cutting. Knives, scissors, saws, and axes can all be used to cut, chop, and slice through fabric, rope, wood, and even steel. Each of these tools is an example of a device called a **wedge**. Wedges are implements that are wide at one end and gradually taper to a thin edge on the opposite end.

Before people had tools made of metal, they made simple cutting tools from stone, wood, and bone. Archaeologists tell us that the first tools were stone choppers, used for cutting everything from animal hides to vines. In this activity, you will test to see how effective a simple stone chopper is at cutting a piece of rope.

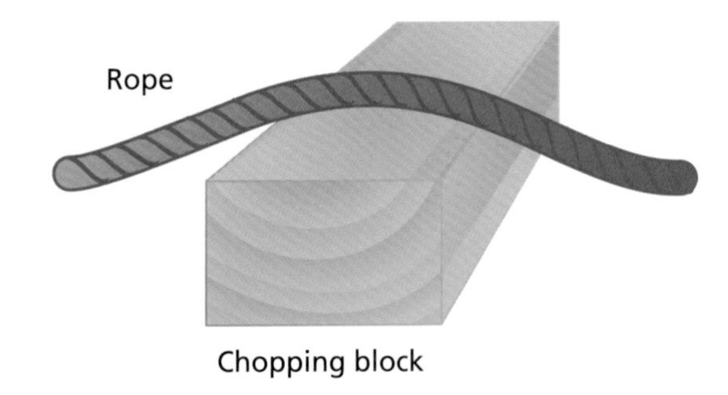

Figure 1

Time Required

45 minutes

Materials

- round rock about the size of a fist
- rock about the size of a fist, with a broken sharp edge
- 12-in.-long (30-cm) piece of 2-in. x 4-in. (5-cm x 10-cm) wood or similar wooden block
- 2 identical pieces of nylon or cotton rope, each about 24 in. (60 cm) long
- safety goggles
- work gloves

Safety Note **This activity requires adult supervision. Make certain that you and anyone near you are wearing goggles and work gloves during this activity. Please review and follow the safety guidelines.**

Procedure

1. Put on the work gloves. Take one piece of rope and grasp it in two hands. Try to rip the rope apart. Record your results.
2. Place the wooden block on a sturdy surface (table or floor). Drape the piece of rope across the middle of the block (see Figure 1). Grasp the round rock in one hand and carefully strike the rope where it crosses the middle of the block. Do this 20 times. Make certain that when you hit the rope with the rock, you are hitting it in the same place each time. Be careful not to hit your fingers with the rock.
3. Pick up the piece of rope that you just hit with the rock and observe it carefully. What happened to the fibers of the rope? Record your observations. Try to rip the rope apart again and record your results.
4. Take the second piece of rope and grasp it in two hands. Try to rip the rope apart and record your results. Place the second piece of rope on the wooden block like you did with the rope in Step 2. Pick up the rock with the sharp edge and strike the sharp edge against the rope 20 times, the same way that you did with the round rock in Step 2.
5. Pick up the piece of rope that you just hit with the sharp rock and observe it carefully. What happened to the fibers of the rope? Record your observations and then try ripping the rope apart. Record your results.

Analysis

1. What was the effect of hitting the rope with the round rock? Why?
2. What was the effect of hitting the rope with the sharp rock? Why?
3. Based on the results of the experiment, how might you improve the cutting ability of the second rock?

What's Going On?

In order for an object to be an effective cutting tool, at least one edge has to be wedge-shaped. The sharper the edge, the easier it will cut. Modern cutting tools have extremely sharp edges, which allow them to easily split and separate the fibers of the material being cut. Early humans discovered that rocks with natural wedge shapes were better than rounded rocks at cutting and splitting materials. Rounded rocks could break fibers by smashing them. Wedge-shaped rocks would split the fibers apart, making it easier to separate them.

Our Findings

1. Hitting the rope with a round rock will cause some of the fibers to break, but it will probably not break enough of them to allow you to tear the rope apart.
2. Hitting the rope with the sharp edge will cause the fibers to split and tear. This is because the wedge-shaped edge cuts through the fibers of the rope.
3. If you wanted to improve on the cutting ability of the sharp stone, you could use another rock to chip away at the edge so that it is smoother and more tapered. The more gentle the taper, the sharper the cutting tool.

WORKING WITH WEDGES

Although simple stone choppers helped early humans carry out many tasks, their cutting ability was still limited. Through trial and error, people eventually discovered that they could make a better tool by sharpening both sides of the cutting face. They did this by using another rock called a "hammer stone" to gradually chip away at the edge of the chopper. This led to the development of a more sophisticated tool known as a hand axe. Unlike a simple chopper, the hand axe had a much narrower blade that was better for cutting.

Both the chopper and hand axe are examples of wedges. Many common woodworking tools used today are simply variations on a basic wedge design. Depending on the shape of the wedge, tools can be used for cutting, splitting, or shaving wood. In **Experiment 1: *The Wedge Design of a Stone Chopper***, we saw how a wedge could be used for cutting. In **Experiment 2: *How a Tool's Wedge Shape Affects Wood Split***, we'll examine which type of wedge is best at splitting a piece of wood.

A hatchet, like an axe, is an example of a wedge, a tool that is wide at one end and narrows to a point or edge at the opposite end. It is used for things such as cutting, chopping, and slicing.

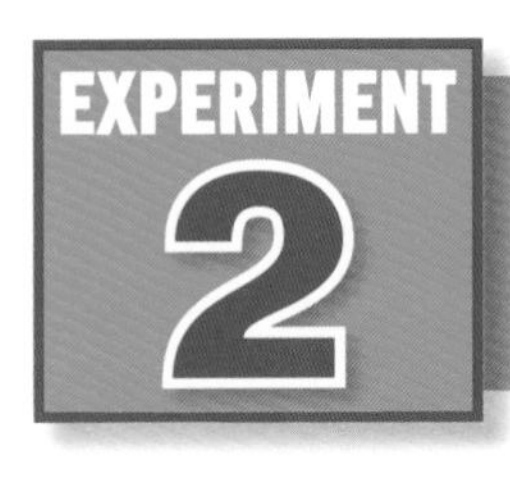

How a Tool's Wedge Shape Affects Wood Split

Topic

Can the shape of a wedge affect how well it can split a piece of wood?

Introduction

Many tools that we use today have a wedge-shaped design. A knife, the blade of a screwdriver, an ax head, and a chisel are all variations on a simple

Figure 1

wedge. Some of these tools are made specifically for cutting. Others are more effective at splitting. The shape of a wedge controls its ability to either cut or split. In this activity, you will test to see how the angle of a wedge controls how effective it is at splitting a piece of wood.

Time Required

45 minutes

Materials

- hammer
- flat-bladed screwdriver
- thick, steel chisel (the type used for splitting brick)
- 2 pieces of 2-in. x 4-in. (5-cm x 10-cm) wood, each about 6 in. (15 cm) long
- safety goggles
- work gloves

Safety Note **This activity requires adult supervision. Make certain that you and anyone near you are wearing goggles and work gloves during this activity. Please review and follow the safety guidelines.**

Procedure

1. Put on the work gloves and goggles. Take the screwdriver and examine it closely. Describe its shape. Predict what will happen when you use the hammer to tap the screwdriver into the wood.
2. Take one piece of wood and lay it on end on a sturdy surface, such as a table or floor. If possible, ask another person to assist you by holding the block of wood steady. Make sure that they put on work gloves and goggles too. Hold the screwdriver straight up and down so that the blade is in the middle of the wood block (see Figure 1).
3. Using the hammer, gently tap the handle of the screwdriver 10 times so that you drive the blade into the wood. Be careful not to hit anyone's

fingers with the hammer. Observe what happens to the wood. Write your observations on the data table.

4. Pick up the chisel and examine it closely. Compare the shape of the chisel to the shape of the screwdriver. Pay close attention to the angle that the wedge makes. Based on what happened with the screwdriver, predict what will happen if you use the hammer to drive the chisel into the wood.
5. Take the second piece of wood and place it on a sturdy surface, as you did with the first block. Place the chisel on the second piece of wood in the same position as you did the screwdriver in Step 2. Tap the chisel with the hammer 10 times and drive it into the wood. Observe what happens to the wood.

Analysis

1. What happened to the wood when you tapped the screwdriver into it?
2. What happened to the wood when you tapped the chisel into it?
3. Which tool required more force to drive it into the wood? Why?
4. If you were splitting logs for a fireplace, what shaped wedge would be the best to use?

What's Going On?

Many woodworking tools are wedge-shaped. The angle of the wedge on each of these tools has a different shape, depending on its purpose. Tools that have a thin blade with only a slight taper, such as a screwdriver, can be driven into a piece of wood without causing the wood to split too much. That's because the angle of the wedge is small and only causes a few wood fibers to break when it enters the block. Tools that have a thick blade, such as the chisel, have a wedge with a much wider angle. As it enters the wood, the wedge forces apart many more wood fibers. This puts a great deal of stress on the block. If enough force is used, the piece of wood eventually will split. Carpenters use a variety of wedge-shaped tools for working with wood. Each tool has its own special shape. A plane has a thin blade that is sharpened on only one side. This makes it ideal for shaving off strips of wood. A maul has a thick blade specifically designed to split a piece of wood. In general, the smaller the angle on the wedge of the tool, the thinner the blade, and the less force needed to use it.

Our Findings

1. When the screwdriver was hit with the hammer, the blade should have gone easily into the wood and the wood should not have split too much.
2. When the chisel was hit with the hammer, there should have been much more resistance and the wood block should have shown signs of splitting, if it did not split completely.
3. Because the chisel has a much wider wedge with a larger angle on the blade, it requires much more force to drive it into the block of wood.
4. If you want to split logs, you would use a very thick wedge with a large angle at the tip.

WEDGES IN THE MODERN WORLD

Today, wedges are used for a variety of jobs. Many doorstops are wedges made of wood or rubber. The doorstop fits between the bottom of the door and the floor. The wedge-shaped design forces the door and floor apart, holding the door open. Wedges also are used for keeping doors closed. Take a close look at the edge of a door where the doorknob is. When you turn a doorknob, a wedge-shaped piece of metal moves in and out of the doorframe. This simple design allows the door to slide closed and lock in place. Wedges also are used in zippers. Most zippers have two rows of teeth that get locked together when the slider passes over them. If you examine the inside of the slider closely, you'll see that there are two wedge-shaped guides. When the slider passes over the teeth, the two wedges line the teeth up and make them mesh together. When you unzip, the wedges pry the teeth apart again.

Wedges don't have to be flat to be useful. A sewing needle, a pin, and an awl are wedges, each with a round shape. The point of a pencil and the tines of a fork also are wedges. When it comes to building things, one of the most important wedges is a simple nail. When you hammer a nail into a piece of wood, the point of the nail forces the wood fibers apart, creating pressure on the board. It is this pressure that keeps the nail stuck in the wood and helps to hold many buildings together.

Of course, before you can nail two pieces of wood together, you have to be able to cut the wood. Today, almost all of the lumber used for buildings and other construction projects is cut using saws. Compared with the hand axe, the saw is a fairly recent invention. The first true saws didn't come about until people started making tools out of metal. Though many of the chipped stone tools used in earlier times had serrated blades, they were not saws. People would have had big problems using these tools to try to cut wood. Most stone tools, like the hand axe, had fairly thick blades with wide-angled wedges. This would have made it difficult for the blade to cut very deeply into the wood. If a person tried to use it to cut back and forth, the wide part of the blade would get stuck in the groove that was being cut. Also, as it cut, a stone blade would leave all the wood fibers in the groove. This would plug up the groove with sawdust, making it impossible to cut further.

For a saw to work properly, it needs a different design. Rather than simply chopping at the wood, the teeth of the saw have to cut and remove wood fibers as the saw slides back and forth. It is possible that the idea for saw teeth came from animal teeth. Some scientists believe that hunters had tried to use the jawbones of animals, such as deer, for cutting through small branches. By running the teeth over the wood, they could gradually break through the fibers, making a relatively straight cut. The first metal saw looked a lot like a large kitchen knife with a row of wedge-shaped teeth along the cutting edge. In **Experiment 3: *How a Saw's Teeth Impact Cutting*,** you'll discover for yourself the unique design that makes a saw so effective for cutting wood.

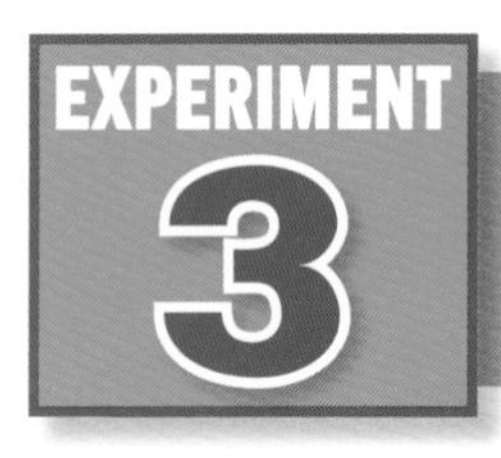

How a Saw's Teeth Impact Cutting

Topic

Do the size and shape of a saw's teeth affect how well it can cut a piece of wood?

Introduction

Based on the archaeological record, the first true saws were used about 4,000 years ago. This is a time when people first began fashioning tools from copper. Drawings from ancient Egypt dating back to about 1,500 B.C. show workers using a small saw to cut wood. The curved metal blade had a serrated edge and was attached to a wooden handle. These early saws were about the size of a large butcher knife, and they cut by ripping across the surface of the wood. In this activity, you will test to see why a saw blade can effectively cut through a piece of wood and see how the size of the teeth controls the way the blade cuts.

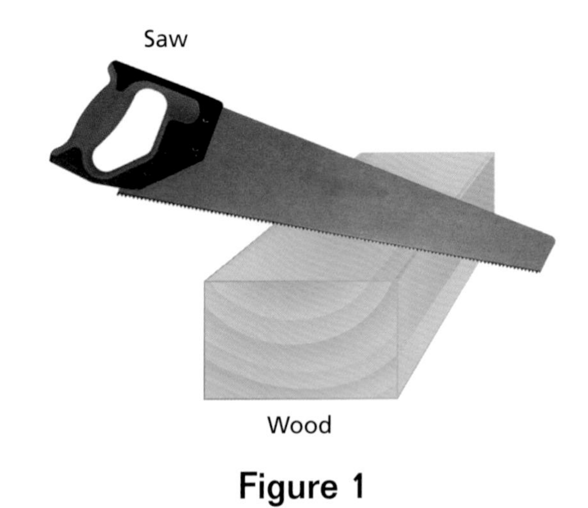

Figure 1

Time Required

45 minutes

Materials

- crosscut hand saw for wood with large teeth
- hack saw or coping saw with small teeth
- metric ruler
- 12 in. (30 cm) piece of 2-in. x 4-in. (5-cm x 10-cm) wood or similar sized wooden board
- safety goggles
- work gloves

Safety Note **This activity requires adult supervision. Make certain that you and anyone near you are wearing goggles and work gloves during this activity. Please review and follow the safety guidelines.**

Procedure

1. Put on the work gloves and goggles. Pick up the saw with the small teeth (hack saw) and examine the edge of the blade closely. Use the ruler to measure the length of the teeth. Observe the shape of the teeth and record your description.
2. Take one piece of wood and lay it flat on a sturdy surface (table or workbench). If possible, ask another person to assist you by holding the block of wood. Make sure that they put on work gloves and goggles, too. Hold the saw by the handle and begin cutting across the top of the wood. Observe in which direction the saw cuts and what happens to the sawdust as you cut. After making five passes with the saw, remove it from the wood and examine the groove cut into the wood. Compare the width of the groove with the width of the saw blade. Record your observations.
3. Pick up the saw with the large teeth (crosscut saw) and examine the edge of the blade closely. Use the ruler to measure the length of the teeth. Observe the shape of the teeth and record your description.
4. Hold the crosscut saw by the handle and place it on the wooden board so that it is about 3 in. (7.5 cm) away from the cut you made with the hacksaw. Begin cutting across the top of the wood with the crosscut saw.

Observe in which direction the saw cuts and what happens to the sawdust as you cut. After making five passes with the saw, remove it from the wood and examine the groove cut into the wood. Compare the width of the groove with the width of the saw blade. Then compare the width of the first groove you cut with the second groove. Record your observations.

Analysis

1. Based on your observations, in which direction did the hack saw cut, when you pushed it or when you pulled it? In which direction did the crosscut saw cut?
2. How did the width of each groove compare with the width of the saw blade that cut it? Why do you think this was so?
3. How did the width of the two grooves compare with each other? Why do you think this is so?
4. How did the depth of the two grooves compare with each other? Why do you think this is so?
5. Which saw required less force to cut width? Why?

What's Going On?

When a saw cuts into a piece of wood, it makes a groove called a kerf. In order to keep the saw blade from "binding," or getting stuck in the groove as it cuts, the kerf has to be wider then the blade itself. To do this, the teeth of the blade have to be "set" in opposite directions along the length of the blade. If you examine the two blades, you will see that half the teeth are bent to the left, and the other half are bent to the right. As the saw cuts into the wood, it also has to remove the sawdust from the kerf. Otherwise the wood fibers will fill up the groove and make the blade stick. To do this, the teeth on most saw blades are curved slightly so that they rake the sawdust out of the groove as they cut.

Most saw blades cut in only one direction. In general, saw blades that are small and thin like hack saws and copping saws cut the wood on the pull stroke. Saws with large thick blades usually cut on the push stroke. This is because a thin blade will bend more than a thick blade. By cutting on a pull stroke, the thin blade will stay straight as it cuts. If a thin blade cut on the push stroke, it would bend, causing the cut to curve. In addition, because more force is used on the cut stroke, pushing on a thin blade could make it snap.

Generally speaking, the larger the teeth on a saw blade, the more force is required to cut with it. Larger teeth also will result in a wider kerf with a more jagged edge. Blades with small teeth are used for fine detailed cutting, while blades with large teeth are used for coarse cutting.

Our Findings

1. Most hacksaws cut on the pull stroke, while most large cross cut saws cut on the push stroke.
2. In both cases, the width of the groove cut by the saw blade was wider than the blade itself. This is because the teeth of the saw are bent slightly to the right and left along the edge of the blade. This keeps the blade from getting stuck in the groove as the blade cuts down.
3. The groove cut by the saw with the larger teeth was wider because the teeth were longer and more bent.
4. The groove cut by the saw with the larger teeth was deeper because the teeth were longer and removed more fibers as they moved cross the block of wood.
5. The saw with the smaller teeth requires less force to use because the small teeth did not make as large a bite in the wood.

ROCK SAWS AND CUTTING CONCRETE

As it turns out, not all saws have wedge-shaped teeth. Saws used to cut rocks (lapidary saws) and concrete use a different approach. Instead of having little wedges cut and chop at the surface, these blades use an abrasive material to grind their way through the rock. To do this, the abrasive on the blade has to be harder than the rock itself. Two of the more common abrasive materials used today are carbide steel and industrial diamonds. Yet, the rock saw is not a modern invention. In ancient Egypt, people were using a type of abrasive saw to cut through stone, too. Archaeologists believe that the rock saws they used had no teeth. Instead, these devices simply slid back and forth over the surface of the rock. The secret to their cutting power was a layer of wet quartz sand that was placed under the saw blade. Quartz is harder than many of the other minerals that the Egyptians were cutting. Examples of these primitive rock saws have yet to be found, but scientists do have several lines of evidence to show that they were used. Ancient drawings show workers cutting stone with sawlike instruments, and many of the stone coffins found in burial sites have saw marks on their lids.

MACHINES MADE SIMPLE

People usually don't think of something simple, such as a hand axe or a saw, as a machine. In our modern world, the word "machine" is usually reserved for complex mechanical devices driven by engines or motors. It the strictest sense, though, all simple tools are machines: They all help to convert mechanical energy into useful work. To a scientist, the word *work* means moving an object over a distance. Over the years, scientists have come to recognize six basic machine types, all of which are classified as simple machines. We've already discussed one of them: the wedge. The other five are the screw, the inclined plane, the lever, the wheel and axle, and the pulley.

As the term simple machine suggests, these devices are basic. They have few or no moving parts. They are important because they are used in many ways and are often components in much larger, compound machines.

Simple machines work by trading force for distance. If you recall from **Experiment 3: *How a Saw's Teeth Impact Cutting***, a saw with small teeth requires less force to push, but you have to move it many more times, compared with a saw with large teeth. In the next two sections we'll take a look at how all of the simple machines work. First, we'll examine a second type of device put to work by humans. It's the simple machine called the **lever.**

LOOKING AT LEVERS

Like the wedge, a lever is a device that is used in many places. Levers are at work on a teeter-totter (or seesaw) on a playground, and on the handle that flushes a toilet. Early on, humans discovered that a lever could help them accomplish tasks that could not be done with a wedge alone. Take hunting, for instance. Although a simple hand axe was great for cutting and skinning an animal after it had been killed, it wasn't useful for actually hunting game. That's because in order to use a hand axe, you had to be very close to the animal. This didn't work well for fast animals, such as deer, which could simply run away. A hunter could throw a hand axe at a large animal, but even if his aim was accurate, the blade would just bounce off the animal's hide.

The solution turned out to be a simple one. By attaching the blade to the end of a long stick, the spear was born! Scientists aren't sure exactly how long humans have been making and using spears, but based on the evidence, it's believed to be about 100,000 years. Before people started using spears for hunting, they were probably using pointed sticks for digging up roots. A spear is such an effective weapon because it gives a person a type of **mechanical advantage** called leverage. In **Experiment 4: *Lifting with Levers***, we'll dissect how a lever works, so you can see for yourself where leverage comes from.

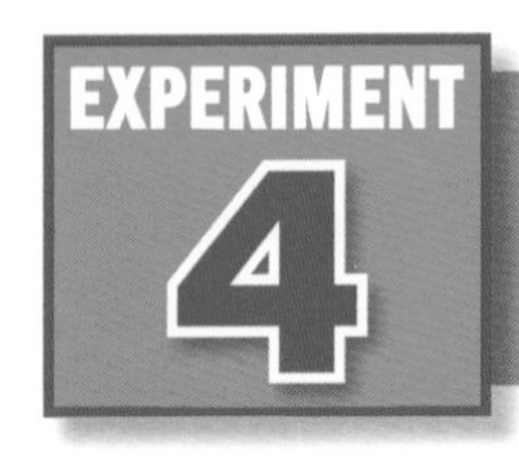

Lifting with Levers

Topic

Can the position of the fulcrum affect the mechanical advantage of a lever?

Introduction

Levers are one type of simple machine. A simple machine is a device that gives the user a mechanical advantage. In science, work is accomplished when an object is moved from one place to another. Every object, from the lightest feather to the heaviest rock, requires a force to make it move. This is called the resistance force. The effort force, as the name suggests, is the amount of force (or effort) used to get the work done. If the effort force is greater than an object's resistance force, then the object will move, and work will be accomplished.

All simple machines reduce the amount of effort force needed to get a job done. They do this by increasing the distance over which the force is applied. The degree that a machine reduces the amount of effort needed is called the mechanical advantage. Simple machines create a mechanical advantage by moving the object a greater distance but with less effort. In the end, the amount of work that is accomplished is the same whether you use a machine or not.

A lever has two parts: a bar and a fulcrum. The bar moves, or pivots, on the fulcrum. The fulcrum divides the bar into two parts called arms (see Figure 1).

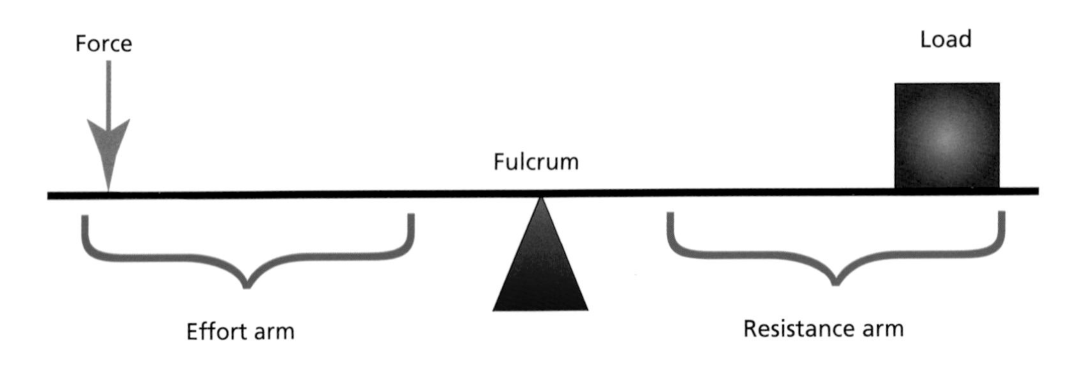

Figure 1

The arm that holds the object being moved (called the load) is the resistance arm. The arm to which you apply the force is called the effort arm. In this activity, you will discover that moving the position of the fulcrum can give you a large mechanical advantage.

Time Required

30 minutes

Materials

- 12-in. (30 cm) ruler
- pencil
- 4 quarters

Safety Note **Please review and follow the safety guidelines.**

Procedure

1. Create a lever by placing a pencil under the 6 in. (15 cm) mark of the ruler. The two ends of the ruler should balance on the pencil. In this setup, the pencil acts as a fulcrum. Place two quarters on the 12-in. (30-cm) end of the ruler and one quarter on the "0" end. The lever should now tip to one side. Record your observations and explain why this happens.
2. Without moving the quarters, change the position of the fulcrum so that the two sides of the lever balance again. Record the number of inches (cm) on the ruler at the point where you get the lever to balance again.
3. Keep the pencil in the new position and predict what will happen if you place a third quarter on top of the two that are already on the 12-in. (30-cm) end of the ruler. Add the third quarter and record your observations.
4. If you wanted to get the two ends of the lever to balance again, predict what you would have to do to the fulcrum. Try to balance the lever again and record your observations.

Analysis

1. Based on your observations, which end of the ruler was the force arm and which was the resistance arm? Why?
2. In which direction did you have to move the fulcrum in order to get the heavier weight to balance with the lighter weight?
3. In order to balance the three quarters, which end of the lever was longer: the resistance arm or the effort arm?
4. Based on your observations, how does moving a fulcrum affect the mechanical advantage of a lever?

What's Going On?

With a lever, the amount of mechanical advantage is controlled by the position of the fulcrum. When the fulcrum is in the center of the bar, the force arm and resistance arm are equal lengths and the mechanical advantage is one. This means that the amount of force required to lift the load is equal to the resisting force of the load. One way to increase the mechanical advantage of a lever is to move the fulcrum closer to the load. This makes the effort arm longer, compared with the resistance arm. When you lifted two quarters with one quarter, you doubled the mechanical advantage of the lever. You did not gain free work, though. You had to move the lever a greater distance. In this case, the effort arm was twice as long as the resistance arm was. When you lifted three quarters with one quarter, you tripled the mechanical advantage, and the length of the resistance arm was tripled.

Our Findings

1. The arm with the multiple quarters is the resistance arm, because it has the load. The arm with the single quarter is the effort arm.
2. In order to lift a heavy weight with a lighter weight, the fulcrum must be moved closer to the load. This increases the length of the effort arm.
3. When lifting three quarters with one quarter, the effort arm was much longer than the resistance arm.
4. With a lever, moving the fulcrum closer to a load increases the length of the effort arm and increases the mechanical advantage.

SPEAR THROWERS AND SIEGE ENGINES

Spears were much better for hunting than simple hand axes. But spears still had a fairly limited range. Also, even the best spears had a hard time penetrating some animal hides. By about 15,000 B.C., hunters had discovered that they could use additional leverage to improve on a simple spear by launching it with a spear thrower. Called an atlatl by the ancient Aztec people, a spear thrower was usually made out of wood, antler, or bone. It looked like a long stick with a curved notch at one end.

A spear thrower extends the length of a hunter's arm, giving him a greater mechanical advantage. The end of the spear is placed in the notch, and the hunter holds the other end of the spear thrower. When the hunter winds up to throw the spear, the longer effort arm (the hunter's arm, plus the length of the spear thrower) means that the spear is released with a greater amount of force. This gives it more power to penetrate tough animal hides. This same type of device is still used by some Inuit today to throw harpoons, and it was the primary hunting tool of many people until being replaced by the bow and arrow.

An aboriginal tribe member in Kimberley, Australia, uses an atlatl. By placing a spear in an atlatl, or spear thrower, a person's arm length increases and this action provides more force in a throw.

During the Middle Ages, levers had a major impact on warfare. Before the cannon was invented, the only way that an army could bring down a stone fortress or castle was to bombard it with large stones from a siege engine. One of the most important of these siege engines was the trebuchet. Thought to originate in China in about A.D. 100, the trebuchet is designed to throw large rocks or other projectiles using a giant lever. A large lever is mounted unevenly on a fulcrum. The short end is the effort arm; it has a large weight (usually a box of rocks) hanging from it. The long end of the lever is the resistance arm. It is attached to a large sling that holds a rock or some other type of projectile. Unlike a catapult that gets its power from some type of spring, a trebuchet uses the movement of the weight to fire the projectile.

A trebuchet uses the movement of weight to fire heavy objects long distances: The weight on the short end of the beam is released, pulling that side down and sending force to the other end of the beam, thereby firing the object.

Before the trebuchet could be fired, a team of men would pull down on the resistance arm, lifting the effort arm and its weight high above the ground. The weight was then locked into place. When the signal to fire was given, the weight was released. As gravity pulled it toward the ground, the force was translated through the fulcrum to the other end of the beam firing the rock. During the fourteenth century, some trebuchets were so large that they could fire a 220-pound (100-kg) rock a distance of 492 feet (150 m). That's nearly the length of two football fields.

LEVERS IN THE MODERN WORLD

Everywhere you turn, you can see tools that are levers in action. Just about any tool with a handle uses a lever to give mechanical advantage. Picks, shovels, rakes, and brooms are examples of devices that gain added leverage by having handles. In **Experiment 5: *Testing Hammer Handle Length***, you can discover how important a handle is for the function of a modern-day tool.

Testing Hammer Handle Length

Topic

Can the length of the handle on a hammer affect how well it can drive a nail into a piece of wood?

Introduction

Archaeologists tell us that about 35,000 years ago, people discovered that **hafting** a stone axe head by putting a handle on it gave them a distinctive advantage over simply holding the stone tool in their hands. Today, we take this simple innovation for granted. Without handles, tools such as hammers, hatchets, and cleavers would be much harder to use. A handle acts like a lever. All levers have a point on which they pivot: the fulcrum. In the case of a hammer handle, the fulcrum is the wrist of the person holding the hammer. If the person changes the position of the handle, he or she changes the location of the fulcrum. This affects how easily the tool will work. In this activity, you will test to see why a handle is an important addition to tools that you strike with and how changing its length affects how easily you get work done.

Time Required

30 minutes

Materials

- hammer
- 3 large nails (10d or larger) that are all the same size
- 3 pieces of 2-in. x 4-in. (5-cm x 10-cm) wood, each about 6 in. (15 cm) long
- ruler
- safety goggles
- work gloves

Safety Note This activity requires adult supervision. Make certain that you and anyone near you are wearing goggles and work gloves during this activity. Please review and follow the safety guidelines.

Procedure

1. Put on the work gloves and goggles. Using the ruler, measure the three nails to make sure they are the same length.
2. Pick up the hammer and hold it so that you are grasping the head directly. The handle should be pointed away from you (see Figure 1). Use your other hand to hold one of the nails steady in the center of one of the wooden blocks. Give the nail five blows with the hammer. Be careful not to hit your fingers. After you have struck the nail five times, use the ruler to measure how much of the nail is sticking out of the wood.

Figure 1

3. Take a second nail and block of wood. This time, hold the hammer by the handle, grasping the handle at the mid point. Give the nail five blows with the hammer. After you have struck the nail five times, use the ruler to measure how much of the nail is sticking out of the wood.
4. Take the last nail and block of wood. Hold the hammer by the handle again, but now grasp the handle near its end. Give the nail five blows with the hammer. After you have struck the nail five times, use the ruler to measure how much of the nail is sticking out of the wood.

Figure 2

Analysis

1. Based on your data, how did changing the length of the hammer handle change the efficiency of the hammer?
2. In which trial did the hammer feel most comfortable in your hand? Why?
3. In which trial did the hammer get the most work done? How do you know?
4. If you were going to chop down a tree, which type of axe handle would you want to use, a long one or a short one? Why?

What's Going On?

When you use a hammer, the handle acts as a lever. A lever has two arms. The effort arm is the end to which you apply the force. The resistance arm is the end at which the force is directed. In the case of the hammer handle, your wrist is the fulcrum, and your own arm is the effort arm. The hammer handle acts as an extension of your own arm. The longer the handle, the greater the force directed to the head of the hammer. Yet, the lever is not providing "free energy." When you swing a hammer with a longer handle, you have to move it a greater distance. That extra distance translates into extra force.

In addition to getting more power out of the hammer, holding a handle toward the end also gives you more control over the tool. This cuts down on wasted motion.

Our Findings

1. In general, a longer handle on a device (such as a hammer) allows the user to gain more power with each stroke. Longer handles make devices more efficient.
2. The comfort when using a tool depends on the user. In general, a tool is most comfortable in a user's hand when it is balanced. So the hammer might feel most comfortable when you are holding it near the mid point of the handle.
3. The most work should have been accomplished in the third trial, when the length of the handle was the longest. The trial data should have shown that the nail was driven into the wood deeper on this trial, compared with the first two trials.
4. An axe with a long handle has a greater mechanical advantage over an axe with a short handle.

CLASSES OF LEVERS

Not all levers are created equally. Scientists have separated levers into three classes, based on where the fulcrum is and how the forces are applied. In a first-class lever, the fulcrum is between the load and the effort. In this type of lever, the effort force and the resistance force go in the same direction. Pliers and a balance scale are examples of first-class levers.

In a second-class lever, the fulcrum is at one end and the load is between the fulcrum and the point at which the effort is applied. The resistance force and the effort force go in opposite directions. A wheelbarrow is an example of a second-class lever; the wheel is the fulcrum.

In a third-class lever, the effort is applied between the fulcrum and the resistance. Like a second-class lever, the resistance force and the effort force work in opposite directions but the point at which the effort is applied is located between the fulcrum and the resistance. The user has to apply an effort force that is greater than the resistance force. The hammer that you used in **Experiment 5: *Testing Hammer Handle Length*** is an example of a third-class lever. The advantage of this type of lever is in the extra control that you gain. One of the most important examples of a third-class lever is the handle that flushes most toilets.

Form and Function in Tools

"Necessity is the mother of invention." This expression is often used to explain why people invent new devices to solve problems. When faced with a particular problem, humans are inspired to invent a new device or process to help solve it. This idea is certainly true when it comes to the development of tools and machines. In fact, it's probably the reason that we have such a variety of tools today. In early times, people had only a few basic tools to help them cut, scrape, hammer, and chop. Hunting created a need for spears and spear throwers. When people learned to grow crops, they needed farm implements. Some of these included sticks for planting seeds, plows for tilling the soil, and sickles for cutting and harvesting grain. Each new tool had a unique form to serve a different function.

RUNNING WITH RAMPS

According to historical record, humans experienced a major change in lifestyle around 10,000 B.C. Instead of simply hunting and gathering wild food, they started growing their own food and raising animals, such as dogs, pigs, and cattle. People invented farming. Instead of having to move around in search of food, people stayed in one place and lived in permanent settlements. Before this time, most shelters that humans built were simple structures that could be put up quickly. Now they could invest more time and effort in building larger, sturdier structures. Huts made from animal skins, grass, and mud were replaced by buildings made of logs, clay bricks, and stone. It wasn't long before the first villages and cities began to spring up.

As buildings got larger, so did the building materials. Moving large stones high off the ground posed a major problem. Levers could be used to lift stones a short distance, but large-scale moving created the need for a different type of simple machine. It wasn't long before the inclined plane was invented. An inclined plane has the same shape as a wedge, but it's used in a different way. This machine is so simple that it has no moving parts. Instead, things are moved up and down an inclined plane. In **Experiment 6: *Amount of Force Needed to Move Objects on Inclined Planes***, you can discover for yourself how an inclined plane makes moving heavy objects a breeze.

EXPERIMENT 6 Amount of Force Needed to Move Objects on Inclined Planes

Topic

Does the angle of an inclined plane affect the amount of force needed to move an object up a ramp?

Introduction

An inclined plane, or ramp, is a simple machine that helps move heavy objects to a higher elevation. As with other simple machines, the person using a ramp gains a mechanical advantage by having to use less force to get work done. Although less force is needed to move an object up a ramp, the object has to be moved a greater distance. The total amount of work done is the same as it would be without a machine.

The angle, or slope, of an inclined plane is called the grade. The grade is the ratio between the height of the ramp and its length. Grade is given as a percent. You can calculate the grade by dividing the height of the ramp by the length of the ramp, and then multiplying by 100. For example, an inclined plane that raises an object 30 feet (10 m) and is 150 feet (50 m) long has a grade of 20% because (30/150 x 100 = 20) and (10/50 x 100 = 20). An inclined plane

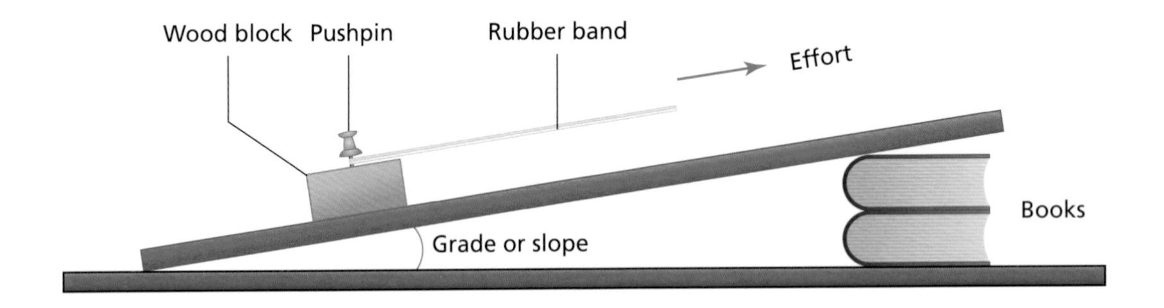

Figure 1

that raises an object 30 feet (10 m) and is 300 feet (100 m) long has a grade of 10% because (30/300 x 100 = 10) and (10/100 x 100 = 10).

In this activity, you will see how changing the grade of an inclined plane affects the amount of force needed to move an object up a ramp.

Time Required

45 minutes

Materials

- 6-in.-long (15-cm) piece of 2-in. x 4-in. (5-cm x 10-cm) wood or similar sized wooden block
- large thumbtack or pushpin
- pen or fine point marker
- ruler
- several thick books to make a stack 18 in. (45 cm) high
- flat board about 4 in. (10 cm) wide and 30 in. (75 cm) long
- large, thick rubber band that has been cut to make one long strip
- friend to assist you

Safety Note **Please review and follow the safety guidelines.**

Procedure

1. Lay the strip of rubber band flat on a table. Without stretching it, measure in ½ in. (approx. 1 cm) from each end and mark the points with the pen. Measure in ½ in. (approx. 1 cm) from one end of the wooden block and mark the point with the pen. Place the mark on one end of the rubber band on top of the mark on the block. Attach the rubber band to the wooden block by pushing the thumbtack through the rubber band into the block at the points you've marked.

2. Grasp the free end of the rubber band in your fingers at the mark you made. Slowly lift the block off the table using the rubber band. The rubber band will stretch. Have your assistant measure how far the rubber band stretches. Record this number on the data table under Trial #1.
3. Construct the ramp on a floor or sturdy table by stacking up the books on top of one another and placing one end of the long board on top of the books. The other end of the board should rest on the floor or table. Using the ruler, measure the height of the ramp and the length of the ramp and record it on the data table in the space for Trial #2.
4. Calculate the grade of the inclined plane by dividing the height of the ramp by the length of the ramp and multiplying by 100. Record this information on the data table under the heading "Grade of Ramp" for Trial #2.
5. Place the wooden block on the bottom of the ramp and grasp the free end of the rubber band in your fingers at the mark you made. Slowly pull the block up the ramp by the rubber band. The rubber band will stretch. Have your assistant measure how far the rubber band stretches when the block first begins to move and record this under Trial #2. Repeat the measurement several times to be sure it is accurate.
6. Remove two or three books from the stack. Reset the ramp. It should be about half the height it was in Trial #2. Measure the height of the ramp and calculate the new grade based on the new height. Record this information on the data table under "Trial #3." Repeat Step 5 and record the length that the rubber band stretches, this time under the heading for Trial #3.

Data Table 1

Trial Number	Ramp Height	Ramp Length	Grade of Ramp	Amount of Stretch
1	X	X	X	
2				
3				

Analysis

1. How did the amount of force needed to lift the block directly off the surface compare with the amount of force needed to lift the block using the ramp?
2. What happened to the grade of the inclined plane as you lowered the height of the ramp?
3. Based on your observations, what happened to the force needed to lift the block as the grade of the ramp got smaller?
4. When using an inclined plane with a smaller grade, what happens to the distance that you have to move the load to get to the same height? How does this affect the amount of work needed to lift the block?

What's Going On?

Inclined planes help people get work done by reducing the amount of force needed to lift an object up to a greater height. In this activity, you measured force by measuring the distance a rubber band was stretched. The more the rubber band stretched, the greater the amount of force. The maximum amount of stretch should have happened when you lifted the wooden block straight off the table. This is because the block was being moved up against the force of gravity, which was pulling it straight down. When you pulled the block up the ramp, the force needed to move it should have been reduced. The ramp was helping to support the mass of the block against the pull of gravity.

When you lowered the height of the ramp, you reduced the grade of the inclined plane. This should have reduced the force needed to move the block up the ramp, compared with a higher ramp. Generally speaking, as the grade of an inclined plane gets lower, it becomes easier to move an object uphill. Using a ramp does reduce the amount of force needed to lift an object, but the total amount of work is the same as if you lifted the object straight off the ground. That's because the ramp requires you to move the object a greater distance. When using a ramp, the total effort expended to lift an object is often greater than the amount of effort needed to lift it directly. That's because there also is friction between the surface of the ramp and the object that is being moved. This extra resistance means more total force must be used, even though the effort needed is less.

Our Findings

1. The amount of force needed to move the block up the ramp (the amount the rubber band stretched) should have been less than when you lifted the block directly off the table. This is because the ramp is supporting some of the mass of the block against the force of gravity.
2. Lowering the height of the ramp reduces the grade of the inclined plane.
3. As the grade of the ramp got smaller, the amount of force needed to pull the block up the ramp also decreased.
4. As the grade of an inclined plane is reduced, the total distance that you have to move an object to raise it to the same height will be increased. The total amount of work stays the same.

INCLINED PLANES IN THE MODERN WORLD

Before people had the use of mechanized equipment, such as bulldozers, dump trucks, and cranes, most of the heavy lifting and moving was done using levers and ramps. Pyramids in Egypt and Mexico were built using ramps to move the large blocks of stone into place. Ramps were also probably used to erect the great stone heads on Easter Island and stone circles such as Stonehenge.

Inclined planes are used in the present day, too. They make up important parts of many modern engineering marvels. For example, almost every major highway has on-ramps and off-ramps, which direct the flow of traffic. Bridges and overpasses are inclined planes. Also, most buildings now have at least one ramp to allow access for people in wheelchairs. Ramps also are used to build staircases. An inclined plane does not have to have a smooth surface. Steps are a type of inclined plane that makes it easier to get from one floor to another. Just think of how much effort it would take to reach

The Egyptian pyramids at Giza and some other ancient structures were built using ramps, which helped people move large stones into place.

the top floor of a building if you had to use a ladder instead of stairs.

SCREWS: RAMPS WITH A TWIST

A screw is a small metal device that fastens things together. Screws are important parts of many construction projects, but they also can be used to move people, cars, and water. If you've ever driven to different levels in a large parking garage or walked to the top of a sports stadium, you've probably noticed that the ramps twist in a spiral. These spiral-shaped ramps maintain a steady slope or grade without taking up a great deal of space. When an inclined plane is wrapped around a cylinder, the result is a screw. In **Experiment 7: *Screw Shape on Inclined Planes***, you will discover how an inclined plane can be turned into a screw and how the grade of the plane affects the shape of the screw.

EXPERIMENT 7 Screw Shape on Inclined Planes

Topic

How does the grade of an inclined plane affect the shape of a screw that is made from it?

Introduction

A screw can be thought of as a modified inclined plane. Like an inclined plane, a screw is a simple machine. Along the edge of a screw are wedge-shaped ridges called *threads*. When screws are used for construction projects, the wedge-shaped threads twist into the material, helping to hold the screws in place. The distance between the top of one thread and the top of the next thread is called the *pitch*. The smaller the pitch, the more closely spaced are the threads of the screw. If two screws have the same length, the one with the smaller pitch will have more threads on it. In general, the smaller the pitch, the more "bite" a screw will have and the better it will hold.

The slope of an inclined plane is called the *grade*. Steeper inclined planes have larger grades. In this activity, you will discover how an inclined plane can be turned into a screw. Then, you will see how the grade of an inclined plane affects the pitch of the screw made from it.

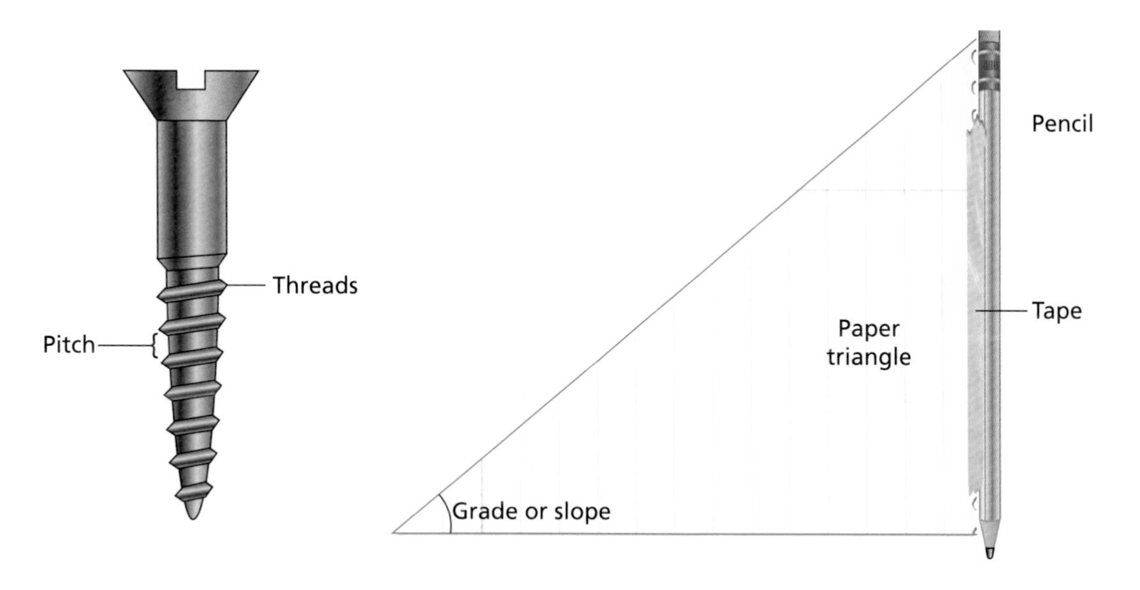

Figure 1

Time Required

45 minutes

Materials

- 2 pencils, each about 6 in. (15 cm) long
- scissors
- cellophane tape
- metric ruler
- 2 pieces of plain white paper, letter size (8 ½ in. x 11 in.)
- felt-tip pen

Safety Note **Please review and follow the safety guidelines.**

Procedure:

1. Take one piece of paper. Using the pencil and the ruler, draw a right triangle that has a bottom length of 5 in. or 10 cm and a height of 2.5 in. or 5 cm. Label this triangle #1. On the second piece of paper, draw a right triangle that has a bottom length of 10 in. or 20 cm and a height of 2.5 in. or 6 cm. Label this triangle #2.
2. Use the scissors to cut out the two triangles outside the lines. The lines should show along the outside edges of the triangles after they have been cut out.
3. Place the two triangles next to each other so that the shortest edge is aligned vertically (straight up and down) along the right hand side. The triangles are the same shape as inclined planes. Compare the slopes of the two inclined planes (triangles). Calculate and record the grade of each inclined plane by dividing the height of the triangle by the length and multiplying by 100.

4. Tape the short edge of triangle #1 to one of the pencils. Tape along the entire length of the pencil so that the paper doesn't come loose (see Figure 1). Slowly turn the pencil so that the paper triangle begins to wind around the pencil. As you wind it, make sure that the line along the bottom edge of the triangle overlaps the line on the pencil from the turn before. After you have wound the entire paper triangle onto the pencil, use another piece of tape to secure the loose end of the triangle to the pencil.
5. Using triangle #2 and the second pencil, repeat Step 4. After you have finished winding the second triangle, place the two pencils side by side and compare them.

Analysis

1. When you wound the triangles onto the pencils, what type of shape did you create?
2. Which inclined plane (triangle) had the steeper grade, #1 or #2? What were the grades of the triangles?
3. Compare the distance between threads on the screw (the pitch of the screw) made from triangle #1 with the pitch of the screw made from triangle #2. Are they different? How?
4. Based on your observations, how does the grade of an inclined plane affect the pitch of the screw that is made from it?
5. If you were screwing these two screws into a piece of wood, which would require more turns in order to go the same depth into the wood?

What's Going On?

A screw is an inclined plane that is twisted around an axis. In this activity, you created two screws using inclined planes with two different grades. In general, the smaller the grade of an inclined plane, the smaller the pitch of the screw made from it. If you have two screws of equal length, the one with the smaller pitch will have more threads. Also, screws with smaller pitches require more turns to reach a certain depth, compared with screws with larger pitches. A screw with a smaller pitch will provide more gripping power, or "bite." This is because there are more threads in contact with the building material.

Our Findings

1. When each triangle was wrapped around a pencil, it made a screw.
2. Inclined plane #1 (triangle #1) had a grade of 50% (2.5 in./5 in. x 100= 50 or 5 cm/10 cm x 100= 50) and inclined plane #2 had a grade of 25% (2.5 in./10 in. x 100= 25 or 5 cm/20 cm x 100= 25)
3. The pitch on the screw made from #1 is larger than pitch #2 because the threads on screw #1 are farther apart.
4. The smaller the grade of an inclined plane, the smaller the pitch of the screw made from it.
5. The screw with the smaller pitch would require more turns to get to the same depth because it has more threads.

THE EVOLUTION OF THE SCREW

As with all simple machines, historians do not know exactly where and when the first screw was invented. Some evidence suggests that screwlike devices came into existence before 300 B.C. But the person who gets most of the credit for putting the screw into practical use was an ancient Greek mathematician and inventor named Archimedes. Born around 285 B.C., Archimedes lived in the city of Syracuse on the island of Sicily. Along with explaining why boats float (Archimedes's principle of buoyancy) and how levers work, Archimedes put the screw to work in a unique way. Instead of using a screw to hold things together, Archimedes designed a screw to pump water.

Instead of making the threads solid wedges, the threads on Archimedes's screw were made from a single hollow tube wrapped around a shaft. The bottom end of the screw was placed under water. As it was turned, water was pushed up the

In Archimedes's screw water pumps, the hollow tube around the shaft pushes water up as the screw gets turned by a crank.

hollow tube to the top of the screw. Though modern technology has given us many new ways to pump water, designs based on Archimedes's screw are still being used in places such as sewage treatment plants. Unlike suction pumps, screw pumps aren't easily clogged by solid material in the water. This is a big advantage when the water being pumped is full of sewage.

SCREWS IN THE MODERN WORLD

Archimedes went on to invent other devices that used screws, including a press to squeeze the oil from olives. Today, screws are at work in many devices. Drill bits are screws designed to cut circular holes in wood, metal, plastic, and even concrete. A vise usually has a screw that allows you to tighten and loosen it. Many tables and chairs have screws on their feet that allow you to adjust their heights. Often, the jacks used to lift cars and trucks each have a large screw running through the center. Though it may not look like it, the propeller that powers a motorboat is really a modified screw, and many machines in factories are driven by special screws called "worm gears."

We really make good use of screws when it comes to keeping the tops on containers. Next time you twist the lid off of a jar of pickles or mayonnaise, look at the inside of the cap. It's really a screw—so are twist-off bottle caps.

By far, the most common use of screws these days is for holding things together. In **Experiment 8: *Fastening with Screws Versus Nails,*** you'll discover why the design of a screw makes a big difference in holding our world together.

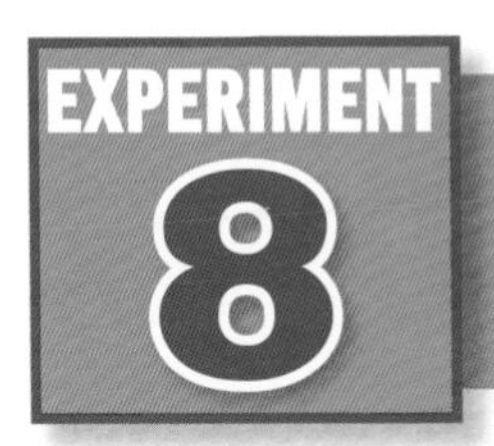

Fastening with Screws Versus Nails

Topic

Does a screw offer any advantages over a nail in fastening materials together?

Introduction

We use many types of fasteners to help hold things together. These include nails, screws, bolts, staples, and tacks. All of these devices fall into one of two categories, based on the type of simple machine after which they are patterned. Nails, tacks, brads, and staples are all variations of a wedge. A wedge is a simple machine that has a point at one end and gradually gets thicker as you move to the other end. When you hammer a nail into a piece of wood, the nail stays in place. That's because the wedge-shaped design forces the wood fibers apart. This creates pressure between the wood and the nail, helping to hold the nail tight.

Many screws have a point at one end, just like a nail. Along the edge of the screw are threads, which form a spiral pattern along the length of the screw. When you put a screw into a piece of wood, the threads cut into the material and grip the wood. Each time you turn a screw, the wedge-shaped threads cut deeper into the wood. Because the screw gets wider toward the top, each successive thread cuts deeper into the wood.

Some screws are the same thickness from top to bottom. Two examples are screws that hold appliances and furniture together and bolts that are threaded like screws but don't have a slot in their heads and have flat bottoms. Instead of cutting into the material being fastened together, these devices are usually held tight by a nut on the other end. Like a screw, a nut is also threaded. When it is placed at the end of a screw, the two threads mesh together. As the nut is tightened, it is held in place by friction.

Time Required

45 minutes

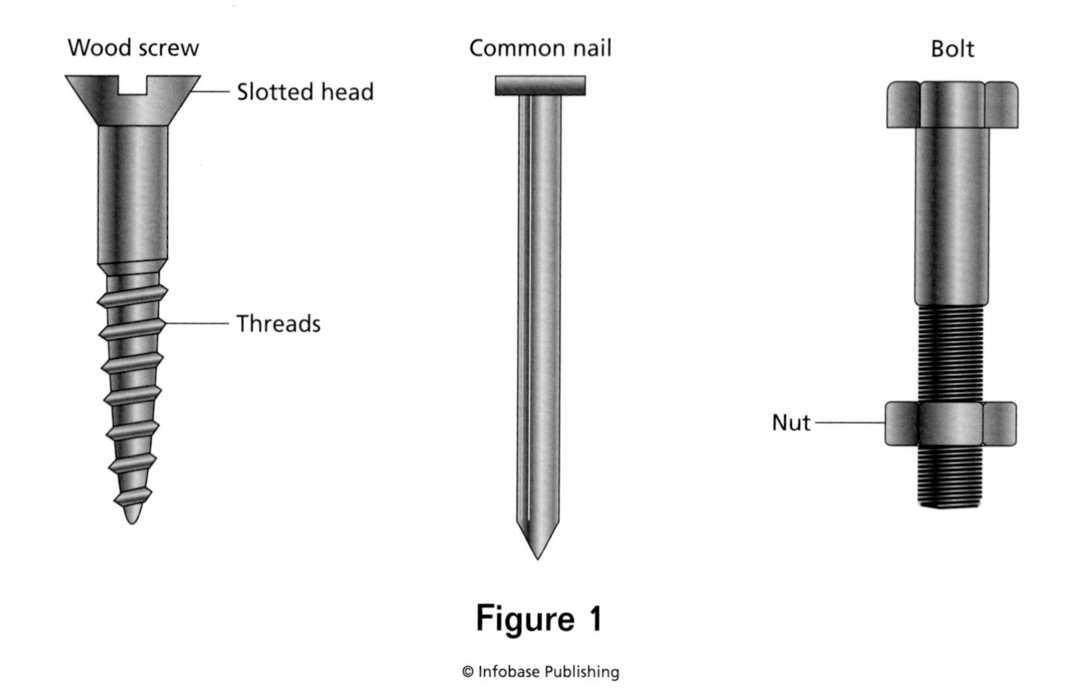

Figure 1

Materials

- goggles
- work gloves
- 6-in.-long (15-cm) piece of 2-in. x 4-in. (5-cm x 10-cm) wood, or a similar sized wooden block
- hammer
- flat-bladed screw driver
- pair of pliers
- ruler
- 2-inch (5-cm) "common" nail
- 2-inch (5-cm) wood screw with slotted head

Safety Note **This activity requires adult supervision. Make certain that you and anyone near you are wearing goggles and work gloves during this activity. Please review and follow the safety guidelines.**

Procedure

1. Place the block of wood on a sturdy, flat surface. Put on the work gloves and goggles. Pick up the nail and place the point in the center of the wood block. Using the hammer, gently tap the head of the nail, driving the nail into the block of wood exactly 0.4 in., or 2⁄5 in. (1 cm). About 1.2 in. (3 cm) of nail should be sticking out of the wood. (You may want to get a helper to hold the block of wood still for you. If you do, make sure that he or she wears goggles and gloves, too.)
2. After you have hammered the nail into the block, examine the way the wood looks around the shaft of the nail. Take the pliers in one hand. Hold the wood block firmly against the table with your other hand. Grasp the head of the nail with the pliers and try to pull the nail out of the wood. If you need to, wiggle the nail back and forth a few times.
3. After removing the nail from the block, turn the block of wood over, so that you have a fresh surface with no holes in it. Place the point of the screw in the center of the block. Using the screwdriver, turn the screw 0.4 in. (1 cm) into the wood. As you turn it, examine the surface of the wood and observe how the screw feels going into the wood.
4. Using the pliers, try to remove the screw from the wood without unscrewing it.

Analysis

1. When you drove the nail into the wood, what did the wood fibers appear to do? How did this compare to what happened to the wood when you put in the screw?
2. Which was easier to remove from the wood, the nail or the screw? Why?
3. If you needed to remove and replace a fastener from a piece of wood several times, which would be better to use, a nail or a screw? Why?
4. Which would be less likely to split a piece of wood, a nail or a screw?

What's Going On?

When fastening pieces of wood together, screws will generally hold tighter than nails because of the way the threads cut into the wood. Both fasteners make holes in the wood, but the shapes of the holes are different. When you hammer in a nail, the wood fibers along the sides of the hole are compressed. The hole created by the nail has smooth sides. The nail is held in the wood by the pressure of the wood against the shaft of the nail.

As you turn a screw into a piece of wood, the wedge-shaped threads cut a groove into the wood that matches the thread of the screw. With each turn of the screw, threads that are already into the wood from the previous turn dig deeper into fresh wood. At the same time, threads that are entering the hole dig into fresh wood fibers along the side of the hole. Because of this, the groove in the hole created by the screw will lock into the thread of the screw. The greater the number of screw threads in the hole, the greater the "bite" of the screw.

Our Findings

1. With the nail, the wood fibers get pushed apart. With the screw, the wood fibers get pushed up and twist around the screw.
2. The nail is generally easier to remove from the wood, because the threads of the screw cut into the sides of the hole in the wood.
3. The screw is a better option if you need to remove and replace a fastener, because you can unscrew it. By twisting it in the opposite direction, you preserve the grooves that were cut into the hole in the wood. Once you remove a nail from a piece of wood, the wood fibers in the hole stay compressed and the hole stays open. If you try to drive the same nail into the same hole, the fibers will not press as hard and the nail will slip out.
4. Both nails and screws can split a piece of wood. With a screw, however, you can drill a small "pilot hole" that will allow the threads to lock into the wood without easily splitting the wood.

DIFFERENT TOOLS FOR DIFFERENT JOBS

At the beginning of this section, we discussed that in most cases, people invented new tools to fill a need. Today, we have so many tools available that sometimes it seems like overkill. Do we really need so many, when a few basic ones can get most jobs done? Let's say that you wanted to take a bent nail out of a piece of wood. Do you really need to use a specialized pair of nail-pulling pliers? There are many other ways to do the same job. You could use a regular pair of pliers, a crow bar, or the claw part of a hammer. In a pinch, even an adjustable wrench can pull out a nail. With so many options available, why did someone go to the trouble of inventing a tool just for pulling nails?

All of these other methods may work some of the time, but they also fail a lot of the time. If a nail is in too deep, regular pliers will not be able to get enough of a grip on it to pull it out. If a nail is bent too much, a claw hammer or crow bar won't be able to get under it without damaging the wood. If the head of a nail is too small or broken off, only nail-pulling pliers will be able to get it out every time. Most people probably don't need these special pliers because they only need to remove a bent nail every once in a while. Yet, for carpenters, who work with hundreds of nails a day, a device that can quickly remove a bent nail saves time and money. In **Experiment 9: *Tool Designs for Tightening***, you will discover how having the correct tool makes a big difference in how fast you can get a job done.

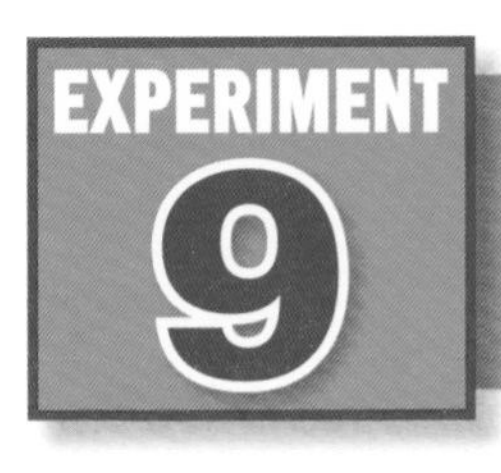

Tool Designs for Tightening

Topic

How does the design of a tool affect your ability to get a job done?

Introduction

We often have a choice of the tool that we can use to get a job done. Different tool designs have different rates of success. In this activity, you will compare the efficiency of two tools for tightening a nut onto a screw.

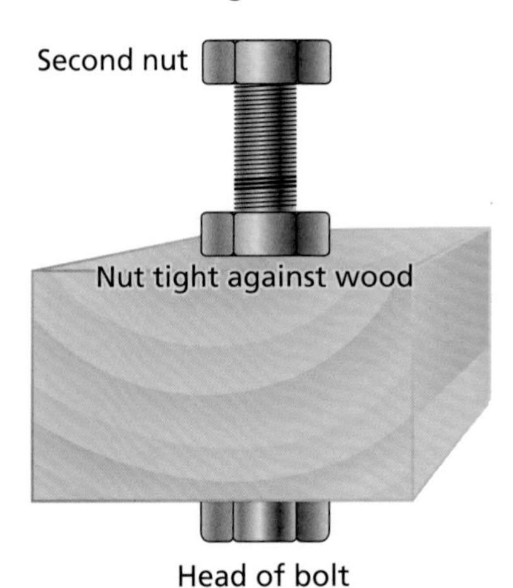

Figure 1

© Infobase Publishing

Time Required

30 minutes

Materials

- goggles
- work gloves
- 6 in.-long (15-cm) piece of 2-in. x 4-in. (5-cm x 10-cm) wood, or a similar-sized wooden block
- machine screw or bolt that is at least 4 in. (10 cm) long
- 2 nuts that fit the threads on the screw or bolt
- watch with second hand
- drill with bit that is wider than the screw or bolt
- pair of pliers
- wrench that exactly fits the nut

Safety Note **This activity requires adult supervision. Make certain that you and anyone near you are wearing goggles and work gloves during this activity. Please review and follow the safety guidelines.**

Procedure

1. Have an adult use the drill to drill a hole completely through the wooden block. Insert the screw (bolt) through the hole in the block so that the head touches one side of the block. The threaded end of the screw (bolt) should be sticking out the other end of the block. Place one nut on the threaded end of the screw (bolt). Using your fingers, spin it down the threads until it is tight against the other side of the wooden block.
2. Take the second nut and place it onto the end of the screw (bolt). Turn it one complete rotation. Using only the pliers, time how long it takes you to tighten the second nut until it comes to rest against the first nut. Record your time.
3. Remove the second nut and reset it at the top of the threads. Turn it one rotation so that you are starting at the same position on the screw (bolt).

Using only the wrench, time how long it takes you to tighten the second nut until it comes to rest against the first nut. Record your time.

Analysis

1. Which tool took more time to tighten the nut, the pliers or the wrench? Why?
2. What problems did you encounter using each tool?
3. How might you improve the efficiency of each tool for doing this task?
4. What other tools might you use for getting this same job done?

What's Going On?

When it comes to tightening a nut on the end of a threaded screw or bolt, both a wrench and a pair of pliers will accomplish the task. However, each has its own advantages and disadvantages. For pliers to work, you must keep squeezing the handles together while turning the nut on the thread. This limits how far you can turn the nut at one time. Because the wrench is designed to fit tightly on the nut, you can turn it farther without stopping. But the opening of the wrench must line up with the edges of the nut. If the two aren't aligned, the wrench will slip off and lose its grip. Also, a wrench can be used only on one size of nut. If the nut is too big or too small, the wrench will not work. There also are adjustable wrenches, which fit a range of nut sizes and shapes.

Our Findings

1. A properly sized wrench will usually tighten a nut faster than a pair of pliers, because the pliers must be taken off the nut and reset more often.
2. The wrench may slip off the nut unless it is properly aligned each time. It is difficult to grip the nut with the pliers and turn it all in one motion.
3. A "box wrench" does not have an open end. It is closed, so that it cannot slip off the nut. "Vise grip" pliers are spring-loaded and lock onto a nut so you do not have to squeeze the two handles together when you turn them.
4. There are a number of tools designed to hold or turn nuts, including a socket wrench, a nut driver, and an adjustable wrench.

INVENTING A BETTER TOOL

It seems like people are always trying to improve on the design of a tool. The next time you are in a hardware or home improvement store, go to the section that has the hand tools and see how many different types of screwdrivers you can find. In a well-stocked store, you might find 100 or more.

At first, this might seem a bit ridiculous. After all, a screwdriver has a really simple job. It turns screws. You shouldn't need more than two screwdrivers because they all work the same way and do the same basic job. Yet, one reason there are so many types of screwdrivers is because there are many types of screws. To turn small screws, you need a screwdriver with a small blade. For large screws, you need a wide blade. If the blade is too small, it will slip instead of grip.

Even when the blade and the screw match in size, slipping can still be a problem when the screw head has a single slot. To solve this problem, screw heads have been developed that allowed a screwdriver to get a better grip. Some screwdrivers have flat blades. Others have blades that look like plus signs (+) and others still look like asterisks or stars (*). These screwdrivers match up with different patterns found on the heads of the screws.

Screwdrivers also differ in their handles. Some have short, fat handles that let you work in places where there is very little space. Others have handles that are long and thin. If you look hard enough, you may find screwdrivers that have handles that are bent at a right angle, or handles that ratchet. These special screwdrivers let you reach screws that are in tight places.

These screwdriver designs didn't come about by accident. They were invented to solve particular problems. Other tools—hammers, saws, chisels, pliers, and wrenches—also come in many styles. In fact, if you can name a tool, you'll probably find that it comes in a range of sizes and shapes, each one designed to meet a different need.

Machines in Motion

So far, we've looked at how simple machines—such as levers, ramps and screws—help get work done more efficiently by reducing the amount of force needed to do work.

When lifting things up and down, gravity is the main force that needs to be overcome. It's not the only force involved in moving things, though. Try this: Place the palms of your hands together and rub them as fast as you can. What do you feel? Chances are, your hands get warm. The warmth comes from another force called **friction.** Friction is the force of resistance that acts between objects when they move against each other. For an object to move, it must first overcome the force of friction.

Let's say someone wanted to slide a book-filled box across a floor. He begins pushing against the box. At first, it doesn't budge. As he applies more force, the box slowly begins to slide. Because he is not lifting the box, all of the force used to move it went to overcome the friction between the box and the floor. Moving a box of books may not require that much effort, but when it comes to moving objects such as refrigerators and freight trains, friction can be a real problem. Not only does it take a great deal of force to get past the friction, but as we found out in the hand-rubbing experiment, friction also creates heat. In **Experiment 10:** ***A Surface's Effect on Friction Between Sliding Objects***, you will test to see if changing the type of surface affects the amount of friction between two objects.

EXPERIMENT 10 A Surface's Effect on Friction Between Sliding Objects

Topic

Does the type of surface affect the amount of friction between sliding objects?

Introduction

Whenever two objects move past and against each other, there is a resisting force between them called friction. The force of friction is always opposite the direction of motion. If you push a chair across the floor to the left, then the force of friction between the floor and the chair legs will be to the right. If an object is placed on a ramp, the force of gravity will tend to pull the object down the ramp. If the object stays at rest on the ramp,

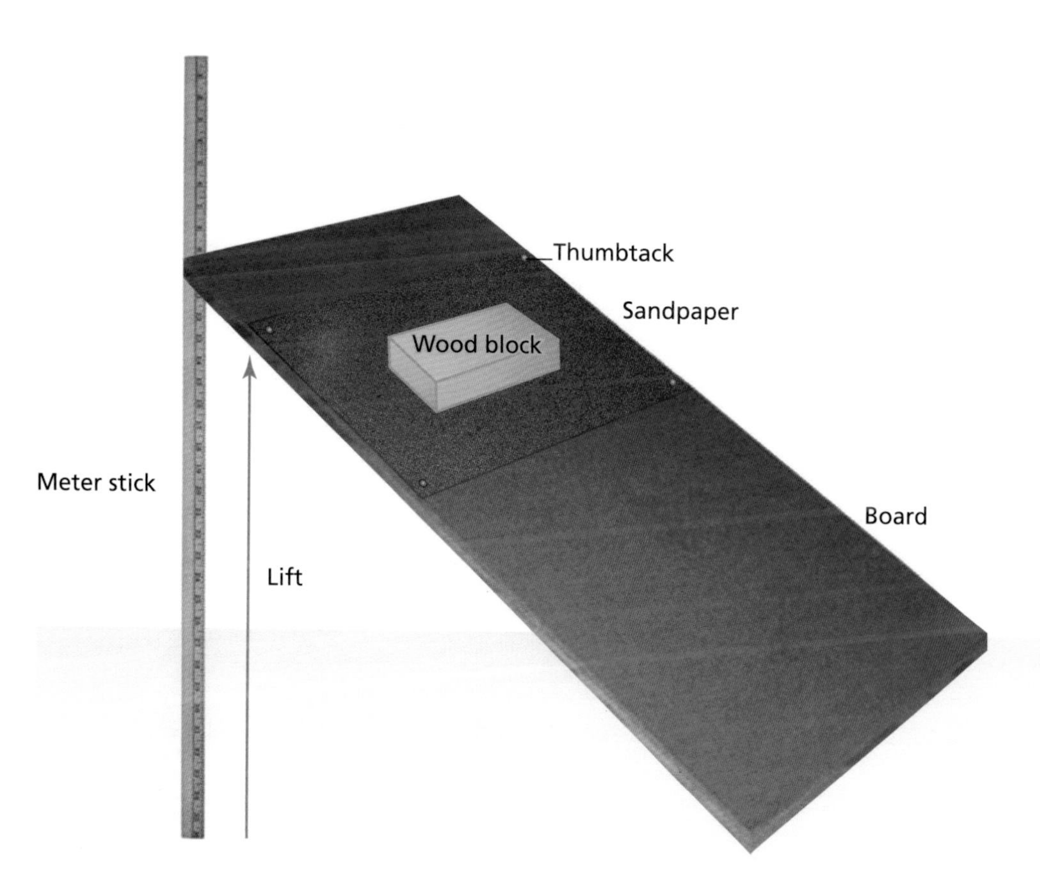

Figure 1

then the force of friction is pushing the object up the ramp with the same force that gravity is pulling it down. If you raise the ramp so that the grade (slope) is steeper, then the downward pull of gravity increases. Eventually, if you keep raising the ramp, the gravitational pull will become so great that the force of friction is overcome, and the object moves down the ramp.

Friction happens because every surface, no matter how smooth, is really covered with tiny bumps and pits. When you try to slide two surfaces past each other, these bumps and pits grab onto each other and "lock up." The force that is used to overcome friction is needed to make the surfaces "unlock." In this activity, you will use a ramp and a sliding wood block to compare the force of friction on several surfaces.

Time Required

45 minutes

Materials

- 6-in.-long (15-cm) piece of 2-in. x 4-in. (5-cm x 10-cm) wood, or similar-sized wooden block
- 4 thumbtacks
- meter stick or large ruler
- flat board at least 4 in. (10 cm) wide and 18 in. (75 cm) long
- 6 in. (15 cm) sheet of sandpaper
- 6 in. (15 cm) sheet of wax paper
- 6 in. (15 cm) sheet of aluminum foil
- masking tape
- friend to assist you

Safety Note Please review and follow the safety guidelines.

Procedure

1. Lay the long wooden board on a flat surface, such as a table or floor. Place the wooden block on the board, on the left-hand side. Place a small piece of masking tape on top of the block. This side will always be the top side of the block. Slowly lift the left side of the board off the surface so that the board forms an inclined plane (ramp) with the lower end to the right. Keep raising the height of the ramp until the wooden block begins to slide down the ramp.
2. Reset the board so that it is again lying flat and the block is on top, on the left side. Repeat Step 1, but this time, ask your friend to hold the ruler next to the left side of the board so that it is pointing straight up. The zero end of the ruler should be resting on the surface next to the board. As you raise the left end of the board, have your friend measure how many centimeters you have raised it at the moment the block begins to slide down the ramp. Record this measurement on the data table. Repeat this two more times, making certain that the side of the block with the tape is always on top. Start the block in the same position on the board each time.
3. After you have recorded three trials with the wooden board, repeat the experiment by changing the surface of the board. Begin with the sandpaper. Attach the sandpaper to the left side of the board using the four thumbtacks. Before conducting the trials, observe the texture of the sandpaper and predict whether there will be more or less friction than with the bare wooden board. Conduct three trials with the sandpaper by placing the wooden block on the sandpaper and lifting the ramp as you did in Steps 1 and 2. Make sure that the block is not resting on the thumbtacks.Record your results on the data table.
4. Remove the sandpaper from the board and repeat Step 3 with the wax paper. Then, repeat Step 3 with the aluminum foil. When attaching the foil and wax paper to the board, don't wrinkle them. Record your predictions and results on the data table.

Data Table 1

Board Covering	Measurement
Wood Trial #1: Trial #2: Trial #3:	

Sandpaper Trial #1: Trial #2: Trial #3:	
Wax paper Trial #1: Trial #2: Trial #3:	
Aluminum foil Trial #1: Trial #2: Trial #3:	

Analysis

1. Based on your results, which surface created the most friction with the wooden block? Which created the least?
2. Why was it always important to have the same side of the wooden block resting on the ramp?
3. Why did you conduct three trials for each surface?
4. Why was it important not to wrinkle the wax paper or aluminum foil?

What's Going On?

In this experiment, you measured the amount of friction between the wood block and the surface of the ramp by lifting one end of the board to form a ramp. The steeper the ramp was, the greater the pull of gravity on the wooden block. The more friction there was, the higher you needed to lift the ramp (the steeper the grade) in order to get the block to slide. Because sandpaper has a rough surface, it causes the greatest amount of friction. Most often, the wax paper has the least amount of friction because the wax makes the paper slippery. Wax is often used as a lubricant to reduce friction.

Our Findings

1. Results will vary, but the sandpaper should produce the most friction because it has the roughest surface. Generally, the wax paper will have the least friction, with the aluminum foil and bare wood falling somewhere in between.
2. You needed to have the same side of the block sliding on the surface so that you kept all the conditions the same in each trial.
3. It was important to conduct three trials so that you would see if you got consistent results each time.
4. Wrinkling the wax paper or foil would have added irregularities to their surfaces, increasing the friction.

MOVING MASSES IN THE ANCIENT WORLD

Modern-day engineers rely on machines—cranes, dump trucks, bulldozers and more—to help them build skyscrapers, bridges, and other large structures. Egyptian engineers building the pyramids some 4,500 years ago didn't have these mechanical helpers. Neither did the builders of Stonehenge, which was erected in England about 4,200 years ago. In both cases, blocks of stone were quarried dozens of miles from the sites of construction. Many weighed more than 80,000 pounds each. So how did those engineers manage to transport such enormous masses over large distances without trucks, trains, or even wheels? According to scientists, the answer lies in some simple physics and a lot of manpower. In **Experiment 11: *Moving Objects with Rollers and Sleds***, you'll have the opportunity to practice some ancient Egyptian engineering and discover for yourself how these monumental tasks were accomplished.

EXPERIMENT 11 Moving Objects with Rollers and Sleds

Topic

Can rollers and sleds make it easier to move heavy objects from one place to another?

Introduction

About 2,575 B.C., construction was started on the Great Pyramid, near what is now the Egyptian city of Giza. This structure, which was the burial tomb of the Pharaoh Khufu (Cheops), is one of the seven wonders of the ancient world. Even by today's standards, it was a massive construction project. It is built from thousands of limestone blocks that were precisely fitted together. Scientists estimate that most of the blocks weigh between 2 tons and 3 tons each, but some of the larger ones weigh more than 10 times this amount. Moving the blocks from the quarry to the construction site posed a tremendous problem.

Most archaeologists agree that the Egyptian engineers didn't use carts with wheels for moving the stones. Instead, they used a combination of sledges and rollers to get the job done. In this activity, you will compare three different techniques for moving heavy objects across a surface to see which is the

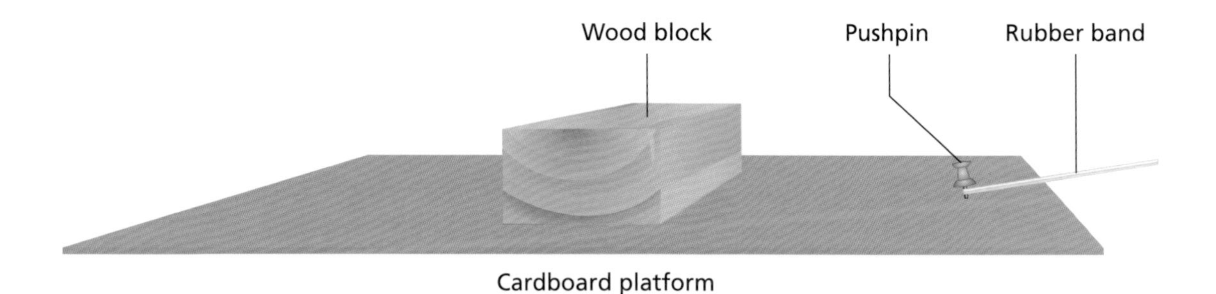

Figure 1

most efficient. Using a "sliding platform" and a rubber band as a gauge, you will test to see which method requires the least force to move the weight.

Time Required

45 minutes

Materials

- 6 in. x 12 in. (15 cm x 30 cm) piece of corrugated cardboard
- 7 pushpins
- 6 unsharpened round pencils
- 2 new, sharpened octagonal pencils
- large, heavy wood block
- large table or desk
- metric ruler
- long, thin rubber band

Safety Note **Please review and follow the safety guidelines.**

Procedure

1. Lay the cardboard on the table in front of you so that the long direction is from left to right. Using the ruler, measure down 3 in. (7.5 cm) from the top of the cardboard and about 1 in. (2.5 cm) from the right side. Mark this point with one of the pencils. Take the rubber band and put it around the point of one of the pushpins. Gently press the pushpin into the cardboard as far as it will go so that one end of the rubber band is held tight against the cardboard. It should look like Figure 1. Test the sliding platform by pulling the cardboard sheet across the table using the rubber band. The rubber band should stretch slightly, but the pushpin should hold tight. If the point of the pushpin pokes through to the other side of the cardboard, place several pieces of tape over it to keep it from scratching the table.

2. Place the block on top of the piece of cardboard. Using the rubber band, pull the sliding platform with the block on top of it across the table in front of you a total of 12 in. (30 cm). Using the ruler, measure and record how far the rubber band stretches as you pull the platform. Repeat the procedure two more times to check your results.
3. Remove the sliding platform and place the six round pencils side by side on the table. Space them about 1 in. (2.5 cm) apart and then place the sliding platform on top of the pencils so the pencils act like rollers (see Figure 2). Using the rubber band, pull the sliding platform on top of the rollers a distance of 12 in. (30 cm), measuring and recording how far the rubber band stretches this time. Repeat the step two more times to check your results.
4. Remove the sliding platform and the rollers from the table. Using the pushpins, attach the two sharpened pencils to the cardboard so they form runners on the bottom. The pencils should be parallel to each other and spaced about 4 in. (10 cm) apart (see Figure 3). You have now turned the sliding platform into a sledge.

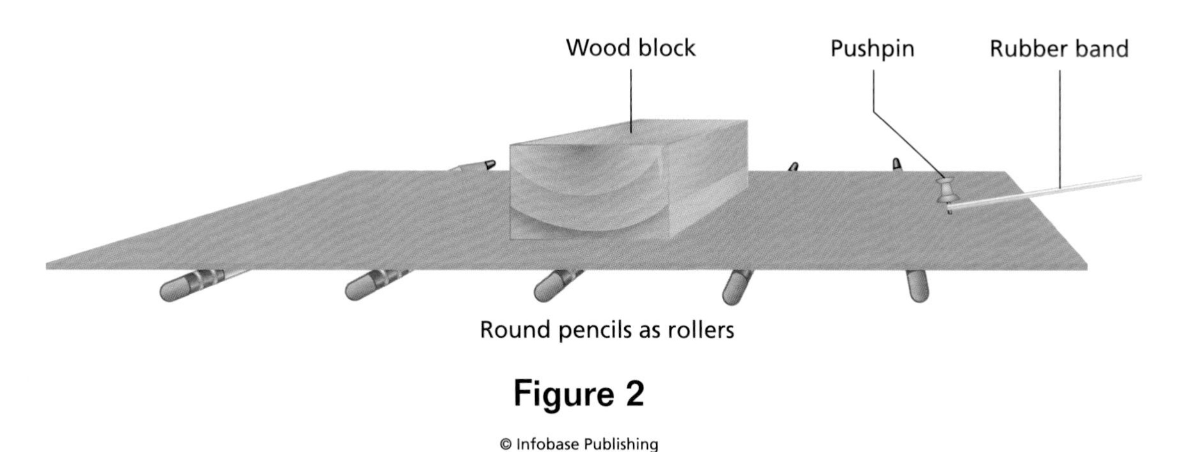

Figure 2

5. Place the sledge on the table with the pencils on the bottom and place the block on top. Using the rubber band, pull the sledge a distance of 12 in. (30 cm), measuring and recording how far the rubber band stretches this time. Repeat the step two more times to check your results.

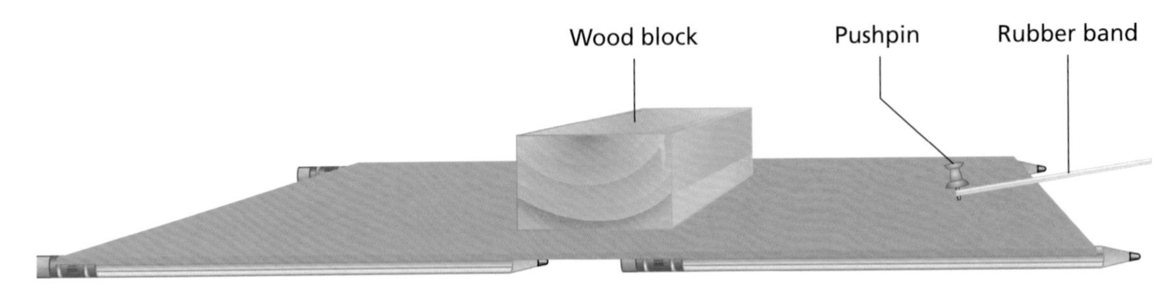

Figure 3

Analysis

1. Which method required the least amount of force to move the block? Which method required the most force to move the block?
2. What were some of the problems that you encountered when you tried using the rollers?
3. What advantage, if any, did having the runners on the sliding platform provide?
4. If you had to move a heavy object over an uneven surface, which method would work the best?

What's Going On?

In this experiment, you measured the amount of force needed to move the block across the table by seeing how far the rubber band stretched. The force required to move the block was controlled by the amount of friction between the bottom of the sliding platform and the table. In general, the greater the amount of surface area between moving objects, the more friction there is. Adding the runners to the bottom of the sliding platform reduces the contact area, so it may help to reduce the friction. The main advantage of having runners, however, is to give the movers more control over the mass. Changing direction and moving over an irregular surface is much easier when you have only two points of contact.

Using rollers reduces the friction the most, but the rollers create other problems. In order to keep a mass moving, rollers from the back have to be moved to the front. This requires a great deal of time and coordination. If the rollers aren't spaced properly, or if they are not set at the proper angle, the mass will either fall off the rollers or change direction unexpectedly. Rollers also work poorly on uneven surfaces because they can turn or get hung up on things. Egyptian engineers probably tried using rollers, but they wound up moving most of the stones with sledges. The sledges were pulled along level pathways. It is also believed that they used water or possibly even olive oil under the runners of the sledges to further reduce the friction and make the runners slide more easily.

Our Findings

1. Using the rollers, the rubber band should have stretched the least, so this method required the least force to move the block. The cardboard sliding platform and the sledge with the runners should have required about the same amount of force to move the block.
2. The rollers had to be kept moving, or the block would have fallen off. The rollers also had to be aligned and spaced correctly.
3. The runners may have reduced the friction a little, but mostly they were useful because they helped to control the direction in which the mass was moved.
4. A sledge would work best on an uneven surface because the runners help to direct the load and even out the rough spots on the surface.

THE AXLE ADVANTAGE

Many people think that the wheel was a natural outgrowth of the roller, but this is probably not the case. Before the wheel was adapted for transportation, it was put to use for other tasks. One of the first recorded uses of a wheel-like device was for shaping clay to make pottery. Potters used these stone "turntables" in the Middle East about 4,000 B.C. Even earlier, people discovered that they could spin fibers of hemp and cotton together to make thread. They used a simple device called a "spindle" to hold the thread. The spindle looked like a top and was a relative of the wheel.

The earliest record of wheels used for transportation comes from the Middle East, about 3,500 B.C. These vehicles were quite simple and looked like sledges with wheels attached to the runners.

Unlike a roller, a wheel can be fixed to an object and still rotate. This is due to the invention of the **axle**. The axle is the device on which the wheel rotates. Without an axle, a wheel would not work.

The development of the wheel and axle was critical for transportation, but these inventions play important roles in many other devices. Together a wheel and an axle are considered a simple machine. This simple machine can reduce the effort needed to get work done. In **Experiment 12: *Testing Wheel Diameter's Effects***, you will test to see how much power can be generated with a simple turn of a wheel.

EXPERIMENT 12 Testing Wheel Diameter's Effects

Topic

How does the diameter of a wheel affect the amount of effort needed to get work done?

Introduction

A wheel and axle can be thought of as a lever in the round. The power of a lever comes from the placement of the fulcrum and the relative length of the lever arms.

With a wheel, the axle takes the place of the fulcrum. By changing the diameter of the wheel, you change the amount of effort required to get work done. Like a lever, a wheel and an axle provide a mechanical advantage: They reduce the amount of effort needed to get work done. In simple machines, moving the object a greater distance creates a mechanical advantage.

A screwdriver is one example of a wheel and axle at work. In this case, the handle of the screwdriver can be thought of as the wheel, and the shaft as the axle. When you turn the handle of a screwdriver, the rotational force from your hand is transferred to the shaft of the screwdriver. This rotational force is called **torque**. Though it may not look like a wheel, a wrench also works like a wheel. When someone uses a wrench to turn a bolt, the bolt rotates. In this

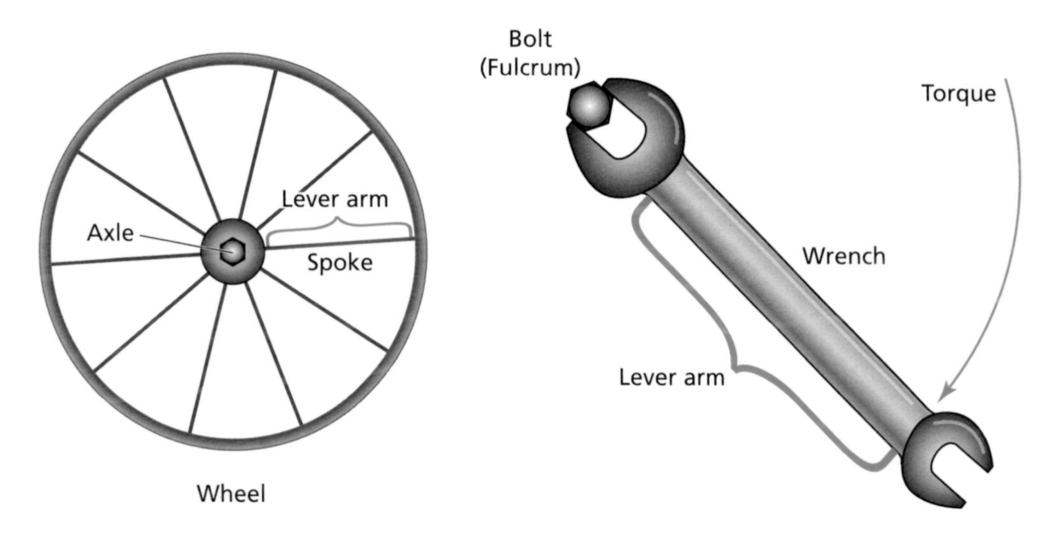

Figure 1

case, the bolt takes the place of the axle, and the wrench would be like one of the spokes of a wheel. In this activity, you will determine how the diameter of a wheel affects the torque by seeing how the length of a wrench affects the ease at which you can make a bolt turn.

Time Required

30 minutes

Materials

- 12-in. (30-cm) piece of 2-in. x 4-in. (5-cm x 10-cm) wood, or similar-sized wooden block
- 2 identical lag bolts, each about 2 in. (5 cm) long (a lag bolt looks like a wood screw with a point on the end of the threads and a head that looks like a nut)
- wrench to fit the head of the lag bolt
- hammer
- 12-in. (30-cm) wooden ruler
- roll of duct tape or electrical tape
- work gloves

Safety Note **Please review and follow the safety guidelines.**

Procedure

1. Put on the work gloves and lay the wooden block flat on a sturdy surface, such as a desk or table. Take the first bolt and place it on the block of wood about 2.8 in. (7 cm) from the left hand side. Using the hammer, gently tap the head of the bolt until the point sticks in the wood. Using the wrench, turn the head of the bolt until it is touching the surface of the block. Notice how far you have to turn the wrench in order to tighten the bolt a half turn.

2. Take the wrench and lay it flat on the table. Place the ruler on top of the wrench so that 4 in. (10 cm) of the ruler is covering the wrench and 8 in. (20 cm) is extending off the back end of the wrench. Use the tape to secure the ruler to the wrench. Make sure that it is tight. The ruler should now act like an extension of the wrench handle.
3. Take the second bolt and place it on the block about 2.8 in. (7 cm) from the right hand side. Using the hammer, gently tap the head of the bolt until the point sticks in the wood. Using the wrench with the ruler attached, turn the head of the bolt until it is touching the wood. Make sure that when you tighten the bolt, you are turning it with the ruler extension. Notice how far you have to turn the wrench in order to tighten the bolt a half turn.

Analysis

1. Which trial required less force to turn the bolt: the wrench alone, or the wrench with the extended handle?
2. Which did you have to turn the greater distance: the wrench alone, or the wrench with the extended handle?
3. If you had to make certain a bolt was tight on a car tire, would you use a wrench with a long handle or a wrench with a short handle? Why?
4. If you had to remove a screw that was stuck tight, would you want a thick handle or a thin handle on your screwdriver? Why?

What's Going On?

When you turn a wrench or a screwdriver, you create a rotational force called torque. Torque is similar to the force created by a lever. With a lever, a longer arm gives a greater mechanical advantage (making it easier to move an object). The same principle applies to the wrench. The longer the wrench, the greater the torque, and the easier it is to turn a nut or bolt. If you turn a wrench one complete rotation, the handle will make an imaginary circle around the nut with a diameter that is twice the length of the wrench. If you were to connect this circle to the axis of rotation, you would have a wheel. This means that the greater the diameter of a wheel, the more torque it produces. The trade-off is that you have to rotate a large wheel a greater distance than a smaller wheel to get the axle to spin the same amount.

Our Findings

1. The wrench with the longer handle should have required less force to turn the bolt.
2. The wrench with the extended handle had to move a greater distance to get the same amount of rotation in the bolt.
3. In order to make a bolt tight, you want a wrench with a long handle. A longer handle gives you more torque, making it easier to tighten the bolt.
4. You want to use a screwdriver with the thickest possible handle to remove a tight screw. A wider handle gives you more torque, making it easier to turn the screw.

WHEELS IN THE MODERN WORLD

It's hard not to find wheels at work in our world today. Almost every form of land transportation has some type of wheel on it. From trucks and trains to skateboards and bicycles, wheels help to reduce friction and make things move more efficiently. Wheels also are used for many other jobs. A winch, used for lifting heavy weights, is a wheel and axle. Many large engines have flywheels inside that keep them working. Cams and cranks are modified wheels, and so are the giant turbines that spin electrical generators in power plants.

You might not realize it, but wheels and axles are all over your home, too. A doorknob is really a wheel and axle, and so is a water faucet. Many tools, including drills, circular saws, belt sanders, lathes, and routers, are based on the design of a rotating wheel and axle. Wheels also are important for sports and play. The handlebars of a bicycle are modified wheels, and so are most tops and yo-yos.

THE POWER OF PULLEYS

Once people discovered the power of the wheel and axle, it didn't take long before they started experimenting with different variations of the machine. One early example that is still in use today is a windlass. A windlass is similar to a winch, with a crank on one end and a large spool in the middle on which a rope is wound. These machines became popular for use above wells to hoist buckets of water and for pulling in sails on boats. Then people discovered that the crank wasn't needed all the time. By simply running a rope over a wheel that had a groove in it, they could lift and move a variety of large objects by pulling on the rope. This simple innovation is called a pulley, and is another one of the six original simple machines.

Historians aren't certain who invented the first pulley, but there is evidence to suggest that the Assyrians were using pulleys in about 600 B.C. One of the first recorded uses was around 390 B.C., when the Greek philosopher Archytas built a number of devices featuring pulleys. But Archimedes really showed the power of the pulley. According to one legend, King Hero, ruler of Syracuse, challenged Archimedes to show how great an engineer he was. Archimedes was more than up to the task. Using an elaborate system of pulleys and levers, Archimedes single-handedly launched a fully loaded ship, complete with crew. In **Experiment 13: *How Pulleys Lift Weight***, you'll discover some of the powerful advantages of pulleys.

How Pulleys Lift Weight

Topic

What are some of the advantages of pulleys when it comes to lifting a weight?

Introduction

A pulley is a simple machine that combines a wheel and axle with a rope, chain, or belt. The rope fits into a groove in the rim of the wheel. By pulling on one end of the rope, you can exert a force on an object attached to the other end. Pulleys are used to change the direction in which a force is applied to an object. They can also be used to change the amount of force needed to make an object move. Pulleys can be moveable or fixed. A fixed pulley is attached to a point that does not move. An example of a fixed pulley would be the pulley at the top of a flagpole. Moveable pulleys are found on many window shades. When you pull the rope on a moveable pulley, the pulley itself changes position, usually moving up and down.

In this activity, you will compare the amount of force needed to lift a block using three methods. Based on your observations, you will rate the effectiveness of a fixed pulley and a moveable pulley for moving a weight.

Time Required

45 minutes

Materials

- 4-in.-long (10-cm) piece of 2-in. x 4-in. (5-cm x 10-cm) wood, or similar-sized wooden block
- 2 nails with large heads, each about 2 in. (5 cm) long
- hammer
- empty thread spool

- wire coat hanger
- wire cutter
- pliers
- work gloves
- goggles
- 40 in. (100-cm) piece of string
- 2 rulers, each 12 in. (30 cm) long
- table, desk, or other sturdy surface
- 2 stacks of books or wooden blocks, each at least 24 in. (60 cm) high
- long, thin rubber band

Safety Note **This activity requires adult supervision. Please review and follow the safety guidelines.**

Procedure

1. Have an adult do this step. Put on the work gloves and goggles. Use the wire cutters to cut the coat hanger and straighten it so that you have one piece of wire about 20 in. (50 cm) long. Put the wire through the hole in the middle of the empty thread spool. Measure 8 in. (20 cm) from one end of the wire. Using the pliers, bend the wire here so that it makes a right angle. Do the same for the other end of the wire, making sure to bend the wire in the same direction on each end. Put the two stacks of books or blocks on a table in front of you so that they are about 6 in. (15 cm) apart. The two stacks should be at an even height. Place one of the rulers between the two stacks like a bridge. Using the pliers, bend the two ends of the wire so they can hook over the ruler. The set-up should look like Figure 1.
2. Carefully hammer one nail into one end of the wooden block so that the head of the nail sticks out about 0.8 in. (2 cm) from the surface of the block. Hammer the second nail in the opposite end of the block so it looks the same as the first nail. Remove the work gloves and goggles.
3. Tie one end of the string to one of the nails and the other end of the string to the rubber band. Slowly pick up the wooden block by lifting up

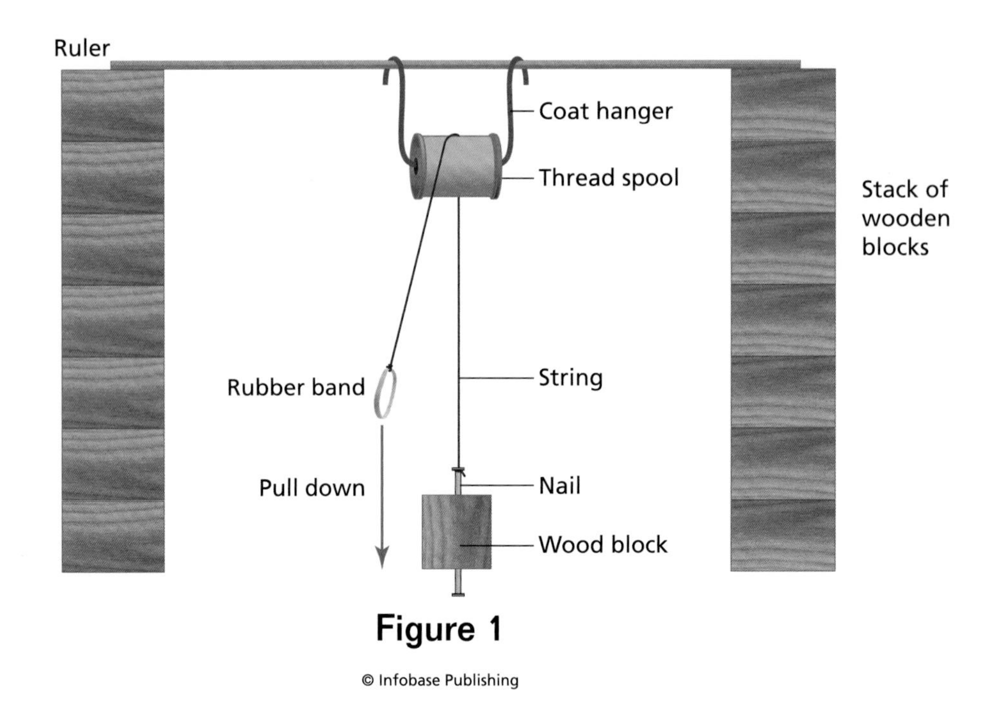

Figure 1

on the rubber band. Lift the block straight up, 12 in. (30 cm) off the table. Using the ruler, measure how much the rubber band stretches when you lift the block. Record the distance on the data table under Trial #1. The amount of stretch equals the amount of force needed to lift the block.

4. Take the end of the string with the rubber band on it and run it over the thread spool between the spool and the ruler from which it is hanging. You are now going to test the effectiveness of a fixed pulley. Pull down on the rubber band so that you lift the block off the table 12 in. (30 cm). Using the ruler, measure how much the rubber band stretches. Record this distance on the data table under Trial #2. Observe how much string you are pulling in order to lift the block 12 in. (30 cm) off the table.
5. Untie the string from the nail on the block. Put on the work gloves and goggles. Using the pliers, unhook the wire and spool from the ruler. Bend the two ends of the wire so that they can hook onto the two nails sticking out of the wooden block. Remove the gloves and goggles and tie the loose end of the string around the middle of the ruler suspended between the two stacks. Push the end of the string with the rubber band on it through the space between the thread spool and the wooden block. The set-up should look like Figure 2.
6. You are now going to test the effectiveness of a moveable pulley. Lift the block and the thread spool 12 in. (30 cm) off the table by pulling up on

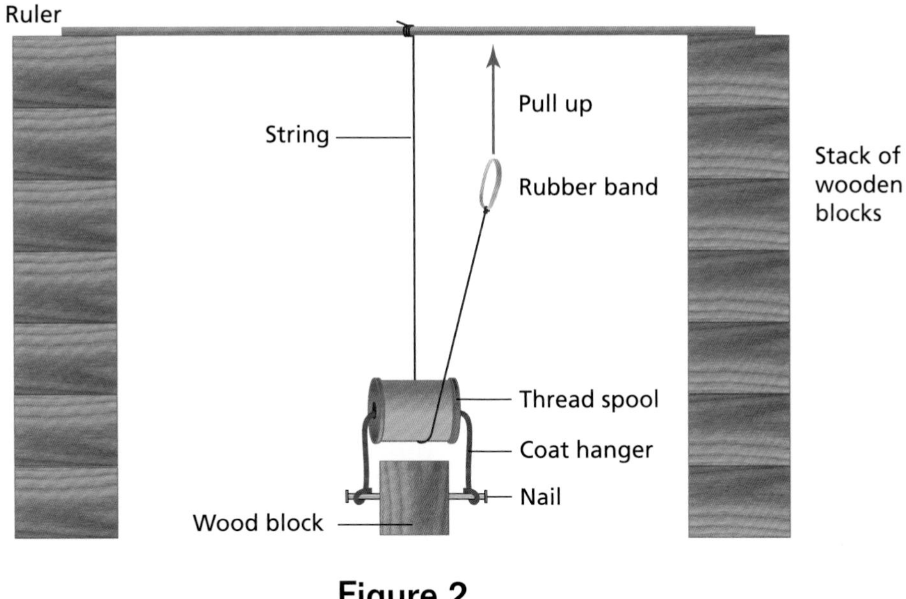

Figure 2

the rubber band. Using the ruler, measure how much the rubber band stretches and record this distance on the data table under Trial #3. While you are lifting the block, observe how much string is being pulled.

Data Table 1

Trials	Distance of Stretch
Trial #1	
Trial #2	
Trial #3	

Analysis

1. Based on the amount the rubber band stretched, how did the amount of force required to lift the block in Trial #1 compare with the amount of force needed to lift the block in Trial #2?
2. How did the amount of force required to lift the block in Trial #1 compare with the amount of force needed to lift the block in Trial #3?
3. How did the length of string you pulled in Trial #2 compare with that in Trial #3?
4. Based on your experiment, what advantage does a fixed pulley have?
5. What are some of the advantages of a moveable pulley?

What's Going On?

In this experiment, you measured the amount of force needed to lift the same mass by using two different pulley arrangements. In Trial #2, you used a single fixed pulley suspended from a bar. A single fixed pulley changes the direction of a force required to move an object, but it does not change the amount of force needed. For example, a pulley at the top of a well allows you to raise a bucket of water by pulling down on a rope. Most people find this motion easier than pulling up. The force required to lift the bucket is the same whether you are pulling up or down; only the direction has changed.

In Trial #3, you used a single moveable pulley attached to the wooden block. In this case, the force required to lift the block was less than lifting the block directly off the table. This reduction in force is called mechanical advantage. To gain this advantage, you have to pull the rope a greater distance. As with all simple machines, the amount of work accomplished by using a pulley stays the same. The trade-off for the reduction in force is that you have to move the object a greater distance.

Our Findings

1. The amount of force required to lift the block without the fixed pulley (Trial #1) should have been about the same as lifting the block with the fixed pulley (Trial #2).
2. The amount of force required to lift the block with the moveable pulley (Trial #3) should have been less than with no pulley (Trial #1).
3. In Trial #3 you needed to pull the string about twice as far as in Trial #2 to lift the block the same distance off the table.

4. A single fixed pulley does not change the amount of force needed to lift an object, but it does change the direction of the force. This is its main advantage.
5. The moveable pulley changes the direction that the force has to be applied and the amount of force needed to lift an object.

PULLEYS IN THE MODERN WORLD

Pulleys often are found in devices used to lift or move objects long distances. Single fixed pulleys can be found at the tops of most flagpoles, at the ends of clotheslines, and on the tops of wells. In these cases, the pulley is used to change the direction of a force, making it easier to move an object.

Single moveable pulleys are usually found on objects that are hoisted up and down repeatedly. Many window shades have these pulleys on each end, as do the booms of sailboats. These pulleys not only change the direction of motion, but also help to reduce the force needed to move an object.

When it comes to heavy lifting, single pulleys usually are replaced with compound pulleys. These devices are important parts of many cranes and elevators. Compound pulleys have multiple wheels through which a single rope or cable is threaded, making multiple loops. Compound pulleys can produce an enormous mechanical advantage. Each time a strand of rope or cable passes through two pulleys, the mechanical

The compound pulley—used by construction cranes—reduces the amount of force needed to lift an object, but the crane's cable needs to be pulled a longer distance in order to lift the object.

advantage is increased. So if you are lifting a 220-lb (100-kg) block with a single moveable pulley, the force needed to lift the block would be cut in half to 110 lbs (50 kg). When using the single pulley, you have to pull the rope 6.65 ft (2 m) for every meter you lift the block off the ground.

If you lift the same block with a compound pulley made of three fixed and two moveable pulleys, there would be five lengths of rope threaded through the pulleys. The effort required to move the block would be one-fifth as much as without any pulleys. This is a greater mechanical advantage, but not less work. In this case, you have to pull the rope 16.6 ft (5 m) for every meter you lift the block. What you gain in reduced effort, you lose in distance.

GEARS: PUTTING SOME TEETH IN WHEELS

Another variation of the wheel and axle is a **gear**. A gear looks like a wheel, but has evenly spaced teeth around its rim. Gears are found in clocks, cars, bicycles, pencil sharpeners, food mixers, and even can openers. Like pulleys, gears can change the direction of motion. They also can be used for changing speed or increasing torque.

It is believed that gears were first used in a device called a noria, which was powered by animals and designed to lift water from lakes and rivers more than 2,300 years ago. About 250 B.C., the Greek philosopher Philon of Byzantium wrote a series of books on simple mechanics. He described several ways in which gears could be put to work. Around 25 B.C., the Roman architect Marcus Vitruvius Pollio showed how gears could be used in combination with a water wheel to power a gristmill. In **Experiment 14: *How Gears Affect Motion***, you will discover how a simple system of gears can be used to change both the speed and direction of motion.

How Gears Affect Motion

Topic

How do gears change the direction and speed of motion in a machine?

Introduction

A gear is a wheel that has teeth along its rim. These interlock with teeth on other gears. Gears are used in machines, such as automobiles, to transfer power from one place to another. In a car, power from the engine is transmitted via a "gear box" to the axles, turning the wheels and making the car go. By using different combinations of gears, the driver can change the speed and the power of the car. In this experiment, you will investigate how the rotations of two gears depend on the number of teeth, or notches, on their rims. In the experiment, gear A has ten teeth and gear B has six teeth. The gears are different sizes to accommodate the different numbers of teeth.

Time Required

45 minutes

Materials

- 2 pieces of 8 in. x 12 in. (20 cm x 30 cm) corrugated cardboard, about 1 in. (2.5 mm) thick
- 2 brass paper fasteners about 1 in. (2.5 mm) long and $\frac{3}{8}$ in. (10 mm) in diameter
- 2 pushpins
- pen or pencil
- scissors
- craft knife or X-acto® knife
- red marker

- 12-in. (30-cm) ruler
- tracing paper or photocopier
- adult to assist you

Safety Note **This activity requires adult supervision. Please review and follow the safety guidelines.**

Procedure

1. Using either the tracing paper or a photocopier, make a copy of the two gears found on the template in Figure 1. Use the scissors to cut out the gears from the copy and then trace around the shapes onto one of the pieces of cardboard.

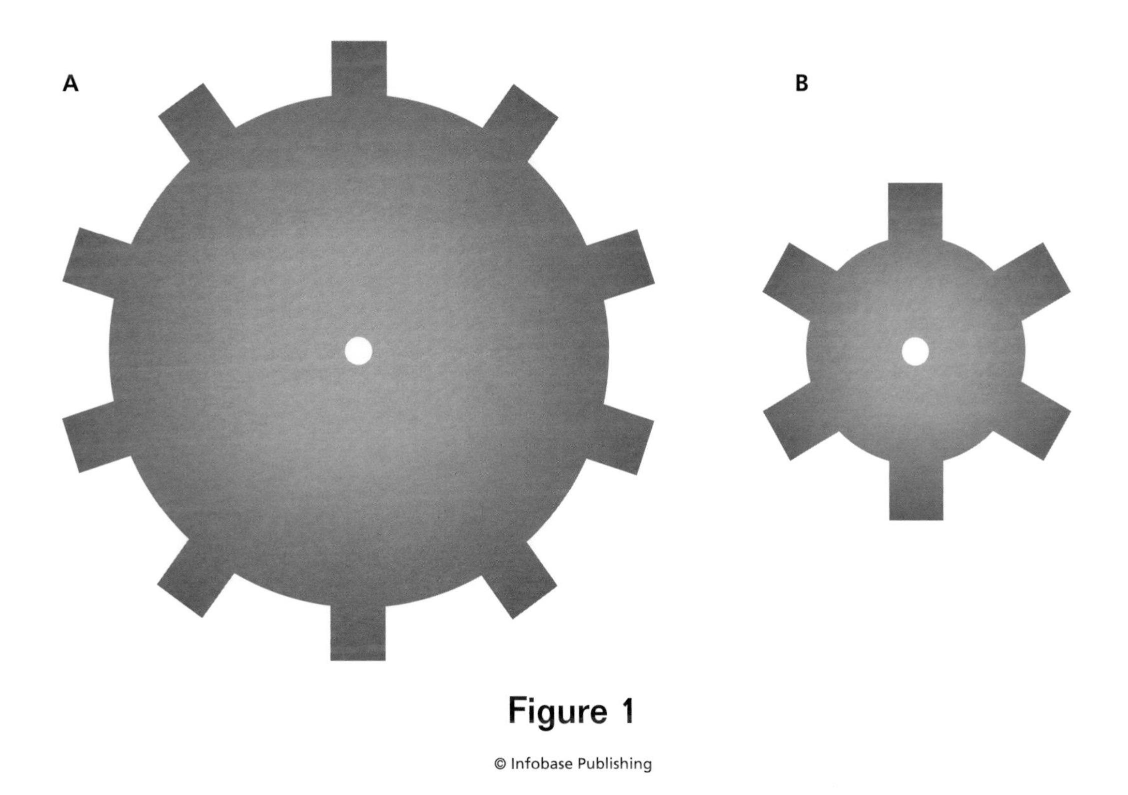

Figure 1

2. Mark one tooth of each gear using the red marker.
3. Ask an adult to cut out the gears from the cardboard using the craft knife (or cut carefully with a sharp scissor). Cut out the notches between the teeth without cutting off the teeth.

4. Label the large gear "A" and the small gear "B". Place the gears next to each other on top of the second piece of cardboard so that they match the arrangement in Figure 2. Push one pushpin through the center of gear A so that the gear is stuck to the cardboard below. The hole must be in the center of the gear or it will not turn properly. Do the same for gear B.

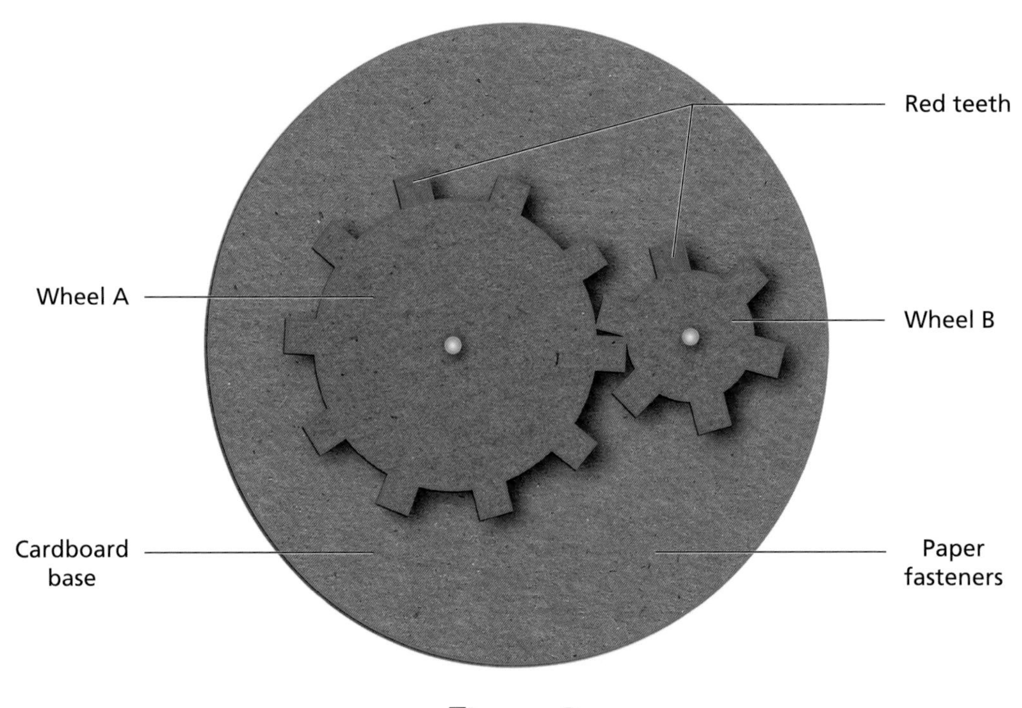

Figure 2

5. Remove the pushpins from the gears. Using the point of the pen, widen the holes in both gears and the piece of cardboard. Place one of the brass paper fasteners through the hole in gear A and then through the hole in the cardboard. Bend the ends of the paper fasteners so that the gear is now attached to the cardboard. Turn it a few times to make sure that it spins freely. If it doesn't, use the pen to widen the holes some more. Do the same for gear B. Make certain that the teeth of the two gears interlock and that the red tooth of each is near the top. This will be the starting position.
6. Begin by using gear A as the "drive gear." It will be the one that you turn with your fingers. Gear B will be the response gear. Turn gear A one full turn clockwise. Record the direction in which gear B moves.

7. Have your partner observe gear B as you turn gear A five complete clockwise rotations. Record the number of rotations that gear B makes.
8. Next, you will use gear B as the drive gear. Predict which direction gear A will turn when you rotate gear B clockwise. Have your partner observe gear A as you turn gear B five times clockwise. Record the direction and how many rotations gear A makes.

Data Table 1	
Action	**Observation**
Moving Gear A	
Moving Gear B	

Analysis:

1. Based on your observations, when two gears mesh, in what direction do they spin relative to each other?
2. When gear A turned gear B, did gear B turn more, less, or the same as gear A?
3. When gear B turned gear A, did gear A turn more, less, or the same as gear B?
4. Based on your observations, when a large gear turns a small gear, what happens to the speed of the smaller gear? When a small gear turns a larger gear, what happens to the speed of the large gear?
5. In order to get two gears spinning in the same direction and at the same speed, what would you need to do?

What's Going On?

When two gears interact, they rotate in opposite directions. If the gears are the same size (have the same number of teeth), they will rotate at the same speed. If a large gear drives a smaller gear, the smaller gear will make more turns for each rotation of the larger gear. This means that the smaller gear is turning faster than the larger gear. When a small gear turns a larger gear, the opposite happens. The large gear makes fewer turns for each rotation of the small gear. This concept is used regularly in cars, motorcycles, and other motorized vehicles. When a change in speed is needed, a device known as a transmission allows the driver to change gears. Changing gears to match the desired speed puts less of a strain on the engine and also helps to save fuel.

Our Findings

1. When one gear turns a second gear, the two gears always spin in opposite directions.
2. Gear B made more rotations than gear A.
3. Gear A made fewer rotations than gear B.
4. When a large gear turns a smaller gear, the smaller gear turns faster. When a small gear turns a larger gear, the larger gear turns slower.
5. In order to get two gears spinning the same direction at the same speed, the two have to be connected at the same time to a third gear.

GEARS IN THE MODERN WORLD

Gears come in a range of shapes and sizes, depending on the jobs for which they are designed. In **Experiment 14: *How Gears Affect Motion***, both gears were flat and they turned in the same plane. These are called "spur" gears. Some machines require that the rotational plane be changed. For example, in an egg-beater, the vertical rotation of the hand crank has to produce a horizontal rotation in the beaters. This is done by using "crown" gears. Crown gears are beveled, or cut at an angle. Another way of changing the rotational plane is by using a "worm" gear. This type of gear looks like a rotating screw in which the threads mesh with the teeth of a spur gear. Finally, gears can also be used to change a rotary motion into a reciprocating or back-and-forth motion. In a "rack and pinion" gear, the pinion is a type of spur gear that rolls back and forth in the "rack," which is a flat track with teeth on it.

One of the important features of gears is that they can be teamed with other components to help power a machine. In a bicycle, for instance, gears are not connected directly to each other. Instead, they are connected with a chain. This allows the power from the pedals to be transmitted to the rear wheel of the bike. In **Experiment 15: *Power Through Gears and a Chain***, you'll discover how the combination of gears, chain, pedals, and wheels allows you the get the most out of your motion.

Power Through Gears and a Chain

Topic

How do a bicycle's gears and chain help to change speed and power as you pedal?

Introduction

Even a one-speed bicycle is a complex machine. A complex or compound machine is a combination of several simple machines. Figure 1 shows the components of a typical bicycle.

When you pedal a bike, the motion of your legs on the pedals is converted to rotary motion by the crank. A pedal is a modified wheel and axle. The crank is attached to a large gear called the pedal sprocket. It turns as you pedal. On bikes with 10 or more speeds, the pedal sprocket has gears of several sizes. Surrounding the axle of the rear wheel is another gear called the wheel sprocket. This is connected to the pedal sprocket with a chain. In multiple-speed bikes, the wheel sprocket can have as many as seven different gears on it. In this experiment, you will determine how the sprockets and rear wheel multiply the power from the pedals, allowing you to travel at a high rate of speed.

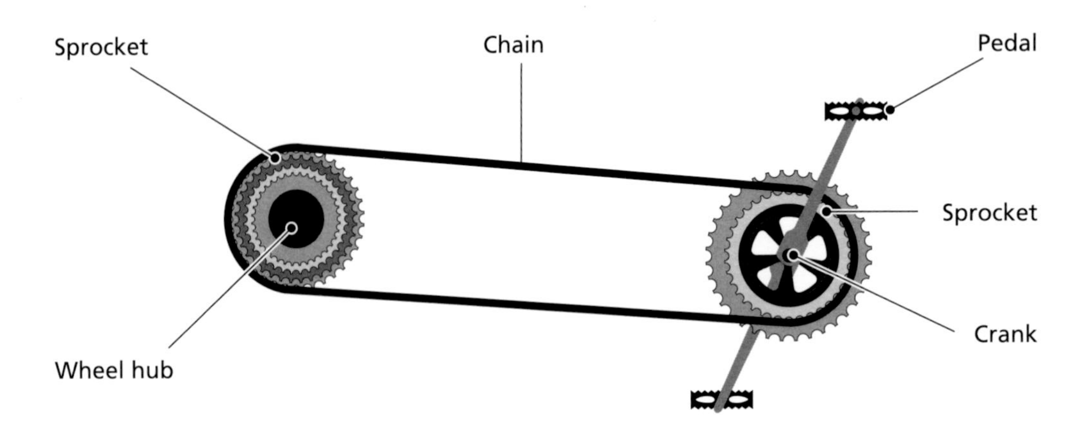

Figure 1

Time Required

30 minutes

Materials

- bicycle
- tape measure or yard stick (meter stick)
- calculator
- pen or pencil
- red permanent marker
- paper towel
- person to assist you

Safety Note This activity requires adult supervision. Please review and follow the safety guidelines.

Procedure

1. Ask an adult to turn the bicycle upside down on the floor so that it is resting on the handlebars and seat. Make sure the handlebars are straight and the bike is stable. Using the paper towel, clean off any excess grease from the wheel sprocket and the pedal sprocket.
2. Select one tooth on each sprocket gear that is pointing straight up or almost straight up. Mark these teeth with the red marker. These will be your reference teeth. If the bicycle has multiple gears on each sprocket, select only one gear on each sprocket to use. Make sure that the chain is connected to the gears that you chose.
3. Starting with the reference tooth, count the number of teeth on the wheel sprocket and record this number on the data sheet. Do the same for the pedal sprocket. Divide the number of teeth on the pedal sprocket by the number of teeth on the wheel sprocket. This number is called the gear ratio. Record the gear ratio on the data sheet.

4. Using one hand, slowly turn the crank on the pedal one complete turn. Observe the direction of motion of the chain and the rear wheel.
5. Allow the rear wheel to stop turning. Set up the two sprockets so that the marked teeth are pointing straight up. Watch the wheel sprocket closely while you turn the pedal sprocket one complete turn. You may want to have someone to assist with this. Count how many times the wheel sprocket turns as the pedal sprocket makes one complete turn. Record this number on the data sheet.
6. Using the tape measure or yard stick, measure the diameter of the rear wheel from one side of the rim to the other, going over the wheel axle. Record this distance on the data sheet. Calculate and record the circumference of the rear wheel by multiplying the wheel diameter by π (3.14).

Data Table 1	
# of teeth on wheel spocket	
# of teeth on pedal sprocket	
# of wheel sprocket teeth divided by # of pedal sprocket teeth	

Date Table 2	
Direction of chain and rear wheel	
Wheel sprocket observations	

Analysis

1. When you turned the crank, which direction did the wheel turn? How is this motion different than if the two sprocket gears were connected directly, without the chain?
2. When you turned the pedal crank one complete turn, how many times did the wheel sprocket turn? How does this number compare to the gear ratio?
3. When you turned the pedal sprocket one complete turn, how far forward did the bicycle move?
4. Based on the results of the experiment, what advantages do the chain drive and rear wheel provide to a person riding a bicycle?

What's Going On?

Chain drives are found on bicycles, motorcycles, some lawn mowers, and chain saws. A chain drive transfers motion from the point of power generation to where it will be used. In the case of a bicycle, the chain connects the pedals to the rear wheel, enabling the rider to propel him- or herself forward. The sprockets found on the pedals and on the rear wheel are gears that change the speed of rotation. Normally, when two gears are connected, they turn in opposite directions. By connecting the gears with a chain, both gears turn in the same direction. This means that when you ride a bicycle, the direction that you pedal is the same as the direction in which the bicycle moves.

When gears of different sizes are linked together, there is a change in speed. When a large gear turns a small gear, the smaller gear will make more turns and the speed will increase. On a bicycle, the pedal sprocket is almost always larger than the wheel sprocket. This means that for every single turn of the pedals, the rear wheel will move more than one turn, which increases the speed of the bike. Dividing the number of teeth on the pedal sprocket by the number of teeth on the wheel sprocket produces a value called the gear ratio. For example, if the pedal sprocket has 48 teeth and the wheel sprocket has 24 teeth, the gear ratio is 48/24, or 2:1. This means that for every turn the pedal makes, the wheel will turn twice. If the wheel sprocket had 12 teeth, then the gear ratio would be 4:1. In this case, the rear wheel would turn four times for every turn of the pedals, resulting in a large increase in speed.

Our Findings

1. The crank and the wheel rotate in the same direction. By contrast, two gears connected together rotate in opposite directions.
2. The wheel should rotate more than the pedals. The number of times the wheel turns for each turn of the pedal should approximately equal the gear ratio. If the gear ratio is 3:1, then the rear wheel will turn three times for every rotation of the pedals.
3. The distance the bicycle moves across the ground is equal to the circumference of the wheel multiplied by the gear ratio. If the bicycle wheel has a diameter of 24 in. (61 cm), its circumference is about 75 in. (190.5 cm). If the gear ratio is 3:1, the bicycle will move 225 in. (571.5 cm) each time the pedal sprocket is turned once.
4. By having a chain drive with two different sized sprockets, a person can travel a large distance by pedaling a small distance. This enables someone to cover a large amount of ground in a short amount of time.

FIGHTING THE EFFECTS OF FRICTION

Wheels, pulleys, gears, and chain drives help to put machines in motion. But they suffer from one major problem: friction. As we discussed at the beginning of this section, whenever objects move against each other, the force of friction causes them to rub together. This rubbing means that extra effort is needed to keep the machines moving. Friction also has another serious side effect: heat. Remember the hand-rubbing experiment? In machines, friction between gears, on wheel axles, and on chains causes a great deal of heat to build up. Often, this heat causes metal parts to expand and lock together, leading to failure.

One way of combating friction is to use a lubricant on the moving parts. If you try rubbing your hands together when they are wet and soapy, you'll probably find that they don't get as warm as when they are dry. This is why car engines need motor oil. The oil coats the moving parts. This not only helps to reduce friction, but also helps to cool the engine. The danger of parts failing due to friction on a bicycle is much less than in the engine of a car. Still, rusty sprockets and chains can make a bike hard to pedal. For this reason, it's important to make sure that the chain on your bicycle and all the gears are clean and well oiled before you ride.

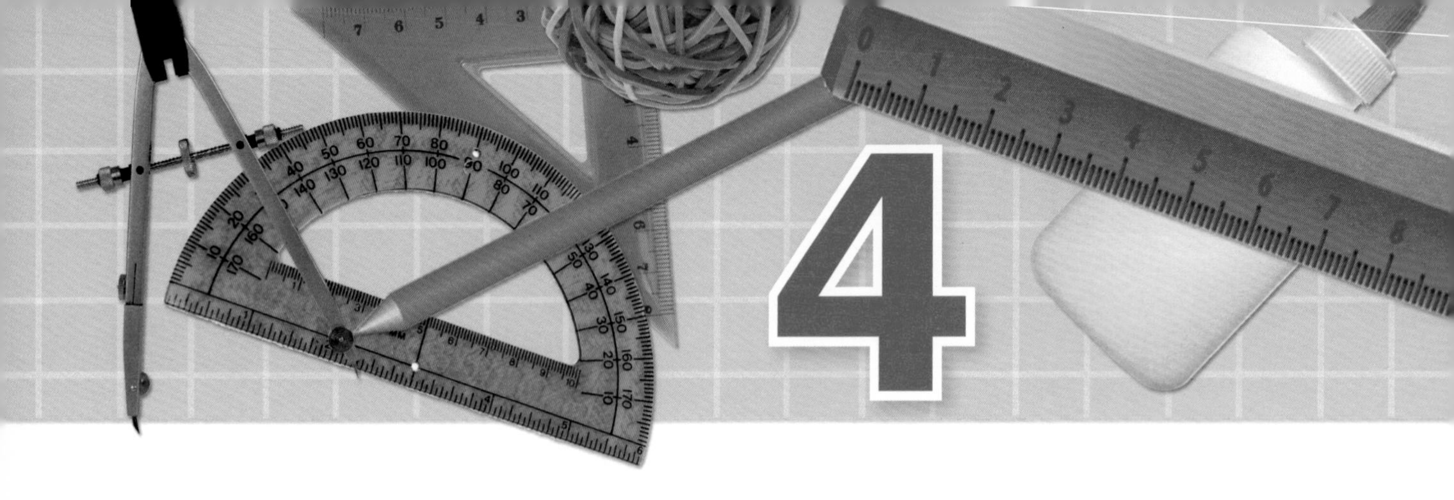

4 Power for the People

With the development of pulleys, gears, and chain drives, people could create bigger and more complex machines. Unfortunately, these new machines led to another problem. They all needed energy sources to make them run. For thousands of years, the only power source was people's muscles. Plowing fields, pumping water, and grinding grain all required more strength than a typical person could muster.

By about 2,000 B.C., people living in Egypt discovered that they could tap into the power of cattle, oxen, and other "beasts of burden" by using rope to tie them to plows and carts. The device that made this possible was a wooden frame called a yoke. At first, cattle were harnessed to the yoke by their horns. This was not very efficient. The yoke was eventually redesigned to fit around an animal's neck. This allowed animals to push against the yoke with their shoulders, producing much more power.

SPRINGING INTO ACTION

Using animals for heavy work was a partial solution to the power problem, but some devices were just not well suited to be powered by animals. People needed a portable power source that could store and release energy on command. The answer came from a simple device called a **spring**.

A spring can be made from any material. The most important property of a spring is that it can be bent or stretched under a force, but will return to its original shape once the force is released. In **Experiment 16: *Testing Spring Materials,*** you will test the spring potential of several materials by building a simple spring-loaded catapult.

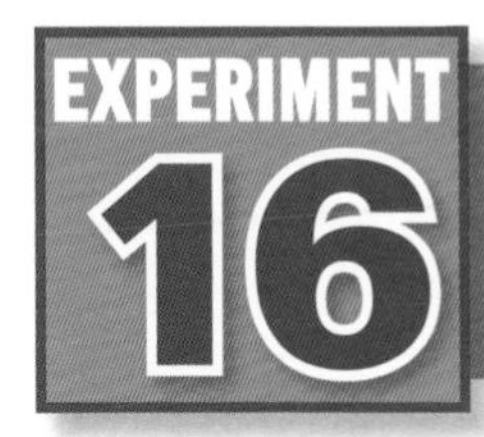

Testing Spring Materials

Topic

Does the type of material from which a spring is made affect its ability to store energy?

Introduction

People usually think of springs as long, stretchy coils made of metal. Coil springs are quite common today, but the first springs were not round and were made of wood. Though it may not look like it, the first spring-loaded device was the bow and arrow. When a bowstring is pulled back, the wood of the bow is bent, storing **potential energy**. When the string is released, the wood returns to its original shape and the string springs forward with **kinetic energy**. Scientists aren't sure when the bow and arrow was invented, but it is believed to date back to around 25,000 B.C.

By 250 B.C., people had begun experimenting with springs made of bronze for use in catapults. In this activity, you will build three simple catapults using springs made out of plastic, wood, and metal to see which is best at storing and releasing energy.

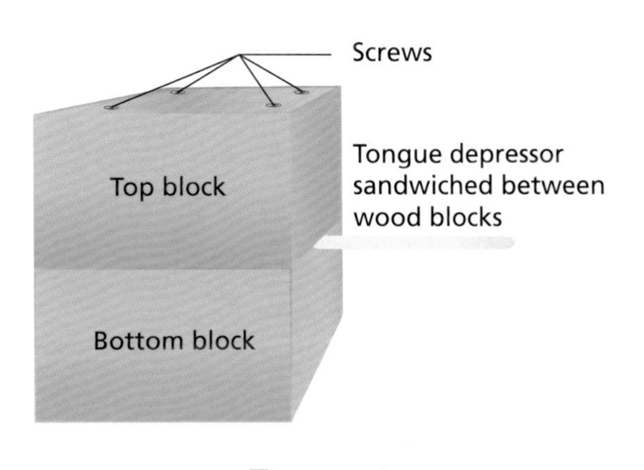

Figure 1

Time Required

45 minutes

Materials

- strip of cardboard 6 in. long x 1 in. wide (15 cm by 2.5 cm)
- disposable plastic knife, about 6 in. (15 cm) long
- wooden tongue depressor or similar-sized piece of wood, 6 in. long x 1 in. wide (15 cm by 2.5 cm)
- 6 wooden blocks, each about 2 in. x 4 in. x 4 in. (4 cm x 10 cm x 10 cm)
- 12 wood screws, each about 3 in. (7.5 cm) long
- screwdriver or power drill with bit to match the screw heads
- ruler
- tape measure
- 1 mini marshmallow
- safety glasses or goggles
- person to assist you

Safety Note **This activity requires adult supervision. Please review and follow the safety guidelines.**

Procedure

1. Build the first catapult, following the design in the figure. Lay one block of wood flat on a table or other sturdy surface. Lay the tongue depressor flat on the block. One end of the tongue depressor should hang over the edge of the block by 4 in. (10 cm). Use the ruler to check the distance. Place the second block flat on top of the first so that the tongue depressor is sandwiched between the blocks. Line up the edges of the blocks and hold them together tightly while an adult screws the blocks together. The four

screws should follow the pattern shown in the drawn figure. The tongue depressor should not slide around at all between the blocks.

2. Following the same procedure you used in Step #1, build two other catapults, first with the cardboard strip and then with the plastic knife. Make certain that each object extends out of the block the same length as the tongue depressor does (4 in. or 10 cm). When using the plastic knife, put the end used for cutting between the wooden blocks and allow the handle to stick out.
3. Place the catapult with the tongue depressor on a hard floor (no carpet) or low table. You will launch the marshmallow straight into the air and your partner will use his or her hand to mark how high it goes. Before you record your results, do one test firing for each catapult to make sure that the springs don't slip. If they do, tighten the screws some more
4. Place the marshmallow at the free end of the tongue depressor. Hold the wooden blocks firmly against the floor with one hand. With one finger of the other hand, press the tongue depressor down so that the tip of it touches the floor. Make sure that your face is not in the way (wear the goggles just in case), and release the tongue depressor. The marshmallow should fly straight up into the air. Have your partner mark the height of the marshmallow with his or her hand. Using the tape measure, measure and record this distance on the data table. Do the test two more times using the catapult made from the tongue depressor. Then repeat the procedure with the other two catapults. Use the same marshmallow for each trial. Record the height of each trial on the data table below.

Date Table 1

Spring Type	**Wood**	**Plastic**	**Cardboard**
Trial #1			
Trial #2			
Trial #3			

Analysis

1. Based on your observations, which spring consistently launched the marshmallow the highest?
2. What properties of the three materials would help to explain these results?
3. Why did you test each catapult three times?
4. Why did you have to make certain that each catapult was the same length and that you used the same marshmallow and launching technique for each trial?

What's Going On?

In order for a material to make a good spring, it has to be rigid enough to keep its shape after being bent, but flexible enough to bend without breaking. This property is called elasticity. All three of the materials you tested in this activity were elastic to some degree, but they produced different results. The cardboard strip bends easily. After a few trials, it does not return to its original shape. The wooden strip bends to a degree, but if bent too far, it snaps. The plastic knife is the most elastic of the three objects and should have given the best results.

All springs work by storing potential energy and then releasing it as kinetic energy. When you push down on a spring, you are using energy. Some of this energy gets stored in the spring as potential energy. When you release the spring, the potential energy gets changed to kinetic energy and the material springs back to its original shape. Materials that bend easily when pressure is applied will not make good springs. Materials that are brittle and break easily under pressure also do not make good springs.

Our Findings

1. The plastic knife should have given the best results.
2. Cardboard is not rigid enough to make a good spring. Wood will work, but it is not as elastic as the hard plastic knife.
3. Tests were conducted three times each so that the results could be checked for consistency.

4. All the catapults were the same length and the same technique was used for each trial in order to keep the number of variables in the experiment to a minimum. If different techniques were used, then the experiment would not be a fair test of the materials.

SPRINGS IN THE MODERN WORLD

Even though there may not be a big need for catapults in our modern world, springs are an important component of many of the machines and devices that we count on every day. Springs come in a range of shapes and sizes.

The most common type of mechanical spring is the "helical" spring. A helical spring usually is made of metal and resembles a large coil wrapped in the shape of a screw. You can usually find helical springs in locks, scales, mattresses, and couch seats. Helical springs keep storm doors closed and garage doors open, and they help return the steering wheel of a car back to its original position after the driver has made a turn. Helical springs also can make great toys, as any Slinky® owner can tell you.

Leaf springs resemble flat metal bars. They provide support for vehicles that carry heavy loads. Most trucks and trains have leaf springs between the axles and the undercarriage. Without these springs to absorb the energy, most large vehicles would break an axle every time they hit a bump.

Before the wide-scale use of electricity, spiral springs were the main sources of power for watches and clocks. Like helical springs, spiral springs are also coiled. They are made from a long, thin, flat piece of metal that looks like a snail shell. When a person winds a watch, they are turning a special gear that tightens a spiral spring. As the spiral spring relaxes, it releases the stored energy at a slow, steady rate.

POWER BY NATURE

The use of springs to power weapons and small mechanical devices rapidly spread through the Roman Empire, but people still didn't have a large-scale power source that could be used to drive mills and pump water. Part of the problem was solved with the development of a wheel-like device called a capstan. When a team of animals was harnessed to a capstan, the animals would walk around in a small circle and the rotary motion could be put to use. This improved things a bit, but didn't solve the problem: Animal-powered machines were still limited by the strength and stamina of the animals. What people really needed was a power source that never got tired. For this, they turned to nature and the forces provided by moving wind and water.

Using wind as a power source was not a new idea. Records show that as early as 3,500 B.C., people were tapping into the wind by building sailboats. Getting wind and water to power pumps and mills was a bit more complicated, however. Around

Primitive norias, which are used to lift liquids, are similar to capstans in that they are powered by animals. Above, the power of a zebu's walk helps at a peanut oil mill in Myanmar (Burma).

200 B.C., the first water mills were constructed based on the idea of a rotating capstan. Instead of being turned by animals walking in a circle, the power came from water that hit paddles attached to a rotating shaft. The whole assembly was placed in or alongside a fast-moving stream. The force of the flowing water made the wheel turn. These early water wheels were a big improvement over animal-powered mills, but they were not very efficient.

The real breakthrough in waterpower came with the development of the vertical waterwheel. On this type of waterwheel, the paddles move up and down like a Ferris wheel, rather than around in a circle like a merry-go-round. Vertical mills are more efficient than horizontal mills, but in order for them to work, they need a system of gears to change the direction of motion. The first record of vertical waterwheels can be traced to the Roman engineer Marcus Vitruvius Pollio. About 25 B.C., he described how to build a mill with a vertical waterwheel. All of the early vertical mills featured "undershot" waterwheels. In this design, the water flowed under the wheel, striking the paddles to make the wheel turn. As with the horizontal mills, the first undershot water wheels depended on natural flowing currents in streams and rivers to make them turn. This meant that when water levels dropped, or when the currents slowed, the mills stopped turning.

By A.D. 500, engineers had figured out how to run mills even during periods of low water flow. Rather than having water flow under the wheel, they created an "overshot" wheel, in which water hit the top of the wheel. This design provided more power than an undershot wheel because the wheel was turned by the flow and by the weight of the water falling on the paddles. Overshot wheels required more work to build. Each one needed an artificial channel to carry the water from the stream to the top of the wheel. Also, in order to keep water levels high all the time, a dam was constructed next to each wheel. The dams created "mill ponds" to store water.

To use waterpower to drive a mill, the mill had to be near moving water. It didn't take long for people to realize that moving air also could be used to power mills. The first record of windmills dates back to A.D. 600 in Persia, which is now Iran. Like an early water mill, an early windmill had a vertical axis and rotated horizontally (like a merry-go-round). The blades were designed to work like sails that captured the wind's energy. The sail design eventually gave way to blades that did the job more efficiently. Waterwheels and windmills are example of devices known as **turbines**. In **Experiment 17: *The Turning Blades of a Turbine***, you will have the opportunity to test the design of a simple wind turbine to see how to get the maximum amount of power from moving air.

EXPERIMENT 17 The Turning Blades of a Turbine

Topic

How does the design of the blades affect the efficiency of a turbine?

Introduction

A turbine resembles a fan. It has a central rotating shaft with blades on it that is designed to spin when the blades are hit by a moving gas or liquid. Turbines can capture the energy from wind, steam, or moving water. They also can be used like an engine to power other machinery. Turbines were first used with waterwheels and windmills. Later, gas-powered and steam-powered turbines were developed. In this activity, you will test to see how the placement of a turbine's blades affects its efficiency.

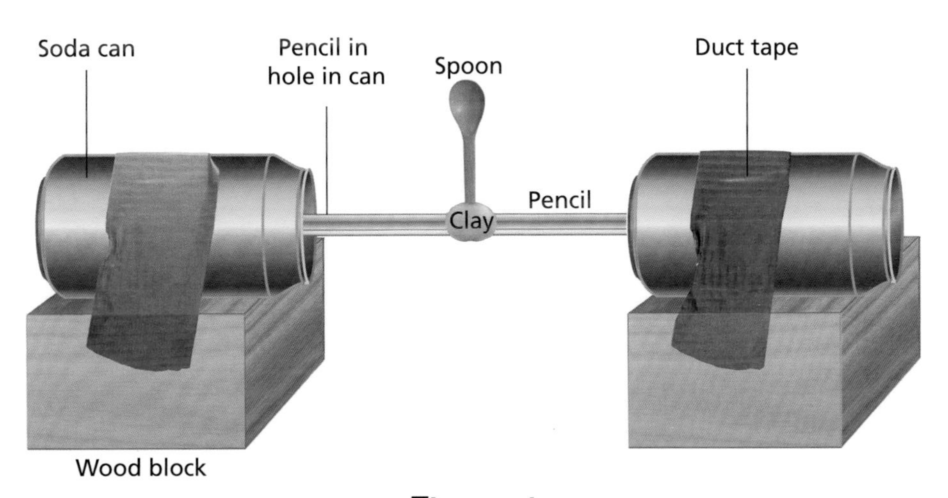

Figure 1

Time Required

45 minutes

Material

- unsharpened wooden pencil—round, if possible
- lump of modeling clay about 1 in. x 1 in. x 2 in. (2.5 cm x 2.5 cm x 5 cm)
- 4 plastic spoons
- goggles
- strong scissors
- ruler
- 2 wooden blocks, each approximately 2 in. x 4 in. x 4 in. (5 cm x 10 cm x 10 cm)
- roll of duct tape or plastic electrical tape
- 2 clean, empty 12-oz soda cans with metal tabs that open them removed
- hand-held hair dryer (blow dryer)

Safety Note **This activity requires adult supervision. Please review and follow the safety guidelines.**

Procedure

1. Ask an adult to use the scissors to cut the handles off the plastic spoons so that about $\frac{4}{5}$ or $\frac{8}{10}$ in. (2 cm) of handle remains attached to the bowl of each spoon.
2. Place the lump of clay on the middle of the pencil shaft. Use your fingers to mold it so that the clay is packed tightly and evenly distributed around the center of the pencil. When you are done, the lump of clay on the pencil should be about 1 in. (2.5 cm) wide. Place one soda can

on top of one of the wooden blocks. Turn the can so its opening is next to an edge of the block. Use the tape to secure the can to the wooden block. Repeat the procedure with the other can and block. The blocks with the cans will be the stand for your turbine.

3. Place the two blocks next to each other on a table. The can openings should be facing each other. Place one end of the pencil in each can, and move the blocks so that the pencil is suspended between the two cans. The distance between the openings of the cans should be about 4 in. (10 cm), and the lump of clay should be centered between the cans. Take one of the cut-off spoon bowls and insert it into the clay so that the bowl is facing you. The spoons will serve as the blades of the turbine. The turbine assembly should look like the drawn figure.
4. Turn the blow dryer on high and point it at the spoon. Observe what happens when the air hits the spoon. Turn off the hair dryer.
5. Turn the shaft of the turbine so that the spoon is pointed straight down. Insert a second spoon bowl into the clay so that it points up. It should be opposite the first spoon. Turn on the blow dryer and point it at the top spoon. Observe what happens when the air hits the spoon.
6. Turn the entire turbine assembly around so that when you are looking at the shaft, the back of the top spoon is facing you. Turn on the blow dryer and point it at the top spoon. Observe what happens when the air hits the spoon and compare the motion to what you observed in Step 5.
7. Turn the turbine assembly around again so that the bowl of the top spoon is again facing you. Turn the shaft so that the two spoons are horizontal. One should be pointing straight at you and the other should be pointed straight back. Insert a third spoon into the clay so that it is pointing straight up. Its bowl should face in the same direction as the bowls of the other spoons. The three spoons should be at right angles to each other. Turn on the blow dryer and point it at one of the spoons. Observe what happens.
8. Insert the fourth spoon opposite the third spoon. The four spoons should be at right angles to one another. Turn on the blow dryer and point it at one of the spoons. Observe what happens. Compare this motion to the other trials.

Analysis

1. What happened when you tried to make the turbine spin with only one blade? Why?
2. Did the turbine turn faster with the bowls of the spoons facing the airflow or with the backs of the spoons facing the air? Why?
3. In which trial did the turbine spin most efficiently? Why?
4. Why is it important for the blades of a turbine to be evenly spaced around the shaft?

What's Going On?

The efficiency of a turbine is controlled by several factors, including the shape of the blades, the number of blades, and the spacing of the blades. On early water wheels, the blades were flat and looked like paddles. This design worked, but people discovered that by curving the blades so that they were slightly concave (like the bowl of spoon), the water would provide more energy. In general, the more blades that a turbine has, the more energy it will get from a moving fluid. There is a practical limit to how many blades a turbine can have. As the number of blades increases, the space between the blades decreases. This interferes with the ability of the fluid to flow past the blades. At some point, it reduces the turbine's efficiency. Before modern turbines are built, engineers often will use computer models to conduct a detailed flow analysis of the fluid as it moves around the blades. In some cases, fewer blades can make a turbine more efficient.

Often, the most important factor controlling the efficiency of a turbine is not the size, shape, or number of blades, but their overall spacing. In order for a turbine to spin efficiently, the weight has to be evenly distributed. This means that the blades must be evenly spaced. If even one blade is off, the turbine will wobble and could fail.

Our Findings

1. The turbine would not spin with one blade because all of the weight was on one side of the shaft. To spin, a turbine needs at least two blades set opposite from each other.
2. The turbine should have spun faster with the bowls of the spoons facing the hair dryer. The concave shape of the spoon bowls traps more moving

air. The backs of the spoons have a convex shape, which causes the air to flow around them.

3. The turbine with four spoons spun the easiest. Increasing the number of blades generally increases the efficiency of a turbine.
4. If the blades are not evenly spaced, the turbine will wobble as it spins, reducing its overall efficiency.

TURBINES IN THE MODERN WORLD

Today, turbines can be found in a range of jobs that require rotary motion. Even though they still follow the same basic principles used by waterwheels and windmills of the past, modern turbines are much more efficient and look very different from these earlier devices. There are four types of turbines. Each type is based on the fluid used to drive it. Water-powered, or hydraulic, turbines are turned by the force of flowing water. They usually are used to generate electricity by turning generators at power plants. Wind-powered turbines also are used to generate electricity. They often can be found clustered in large "wind farms," where several dozen machines turn independently. These turbines are set on large towers that frequently stand more than 300 feet (91 m) tall. The turbines look more like giant house fans than old-style windmills.

Water and wind turbines depend on nature to make them go. Gas turbines and steam turbines—the other two types—rely on heat. Gas turbines can be used for generating electricity too, but most often they are used to power things like airplanes

Wind turbines, like these in the California desert, rely on nature to work by capturing energy from the wind.

and pumps. Most jet engines are powered by gas turbines, which accelerate exhaust gas through a high-pressure nozzle. This provides the thrust to get the plane moving.

When it comes to generating electric power, steam-powered turbines are the real workhorses. These turbines are turned by high-pressure steam created by boiling water. A steam turbine can be combined with many fuels, including coal, oil, natural gas, or even garbage. In nuclear power plants, steam turbines generate the electricity. The nuclear reactor only supplies the heat to turn the water into steam. The turbines do the rest.

TURN ON THE HEAT

Using heat energy to make objects move is not a new idea. It can be traced back to a time when people first discovered what happens to different materials when they get hot. Before we can really understand how heat energy can be put to work to make things move, we must define the word *heat* from a scientific standpoint. When scientists use the word "heat," they are referring to how much internal molecular energy an object has. All matter is made up of tiny particles called molecules. Even though it may not seem like it, all of these molecules are in constant motion, vibrating back and forth. These vibrations are kinetic energy. Heat is a measure of the kinetic energy of individual molecules.

In solids, the molecular vibrations are fairly small. That's why solids keep a fixed shape and size. In liquids, the molecules vibrate faster, so liquids have no set shape. In gases, molecules vibrate very quickly. As a result, gases are free to flow in almost any direction. In **Experiment 18: *How Temperature Change Affects a Gas***, you'll discover how changing the temperature of a gas puts its molecules in motion and how a simple change of state can be used as a source of power.

How Temperature Change Affects a Gas

Topic

How does a change in temperature affect the volume and pressure of a gas?

Introduction

All matter is made of molecules that are in constant motion. When matter is heated, the molecules begin to vibrate faster and the volume of the object changes. These volume changes can then power machines.

Here on Earth there are three common states of matter: solid, liquid, and gas. In some cases, if enough heat is applied, matter in one state (such as a liquid) will change to another state (such as a gas). In this activity you will test to see how the volume of a gas changes when it is heated and cooled, and how changing the state of matter can be used to make things move.

Time Required

30 minutes

Materials

- clean, empty 2-liter soda bottle with labels removed
- round unfilled balloon, 8-inch-diameter (20 cm) or larger
- large, empty mixing bowl (metal) or baking pan
- source of hot water
- watch or timer with second hand
- 15 to 20 ice cubes
- tea kettle with a whistle top filled with about 8 oz (200 ml) of water
- hot plate or stove
- oven mitt

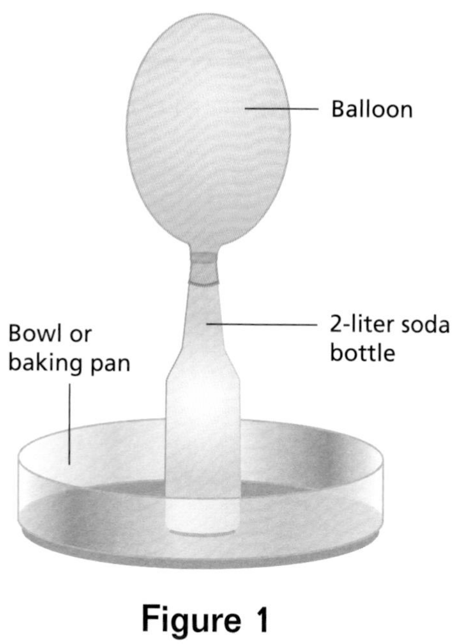

Figure 1

Safety Note **This activity requires adult supervision. Please review and follow the safety guidelines.**

Procedure

1. Place the tea kettle on the stove or hot plate and have an adult turn on the heat. Allow the water to heat for a few minutes and then observe the spout of the kettle. Once the kettle begins to whistle, observe the spout for about 30 seconds, then turn off the heat and allow the kettle to cool for 5 minutes.
2. Place the balloon on top of the empty bottle and place the bottle upright in the empty bowl or baking dish. Observe the balloon to see how much air it has in it.
3. Use the oven mitt to hold the neck of the bottle and ask an adult to fill the bowl or pan about half way with hot water from the kettle. The water should not be boiling, but it will still be hot. Be careful not to splash any on unprotected skin. Allow the bottle to sit in the hot water for one full minute. Do not squeeze the bottle while you are holding it in place. After one minute, remove the bottle and place it on a table. Observe the balloon and note how much air is in it.

4. Allow the bottle to sit on the table at room temperature for three minutes. While you are waiting, empty the hot water from the bowl and place the bowl back on the counter. After three minutes, observe the balloon and note how much air is in it.
5. Place the bottle with the balloon on it back into the empty bowl. Fill the bowl with ice cubes so that the bottle is completely surrounded by ice. Make sure that the ice is packed loosely enough so that it is not physically crushing the bottle. Allow the bottle to sit in the ice for one full minute. After a minute, remove the bottle from the ice and place it on the table. Observe the balloon and note how much air is in it.

Analysis

1. What did you observe coming out of the tea kettle as the water got hot? How did it change once the tea kettle started whistling?
2. What happened to the balloon when you placed the bottle in the hot water?
3. What happened to the balloon when you cooled the bottle with the ice?
4. How could the change of state in the first step and the change in volume in Steps 3 and 5 be used to power a machine?

What's Going On?

Under normal conditions, when most matter is heated, it begins to expand. This is due to the fact that the molecules begin to vibrate faster. The faster they vibrate, the farther they move apart. Gases tend to expand quite rapidly when heated. As soon as a gas is cooled, however, the vibrations of the molecules begin to slow and the gas begins to contract. In a car engine, it is the expansion and contraction of gases due to heating and cooling that drives the pistons to make the engine go.

Gases aren't the only form of matter that expand and contract with temperature changes. If a liquid is heated, it too will begin to expand, but if the heating continues, it will undergo a change of state and become a gas. This change happens because the internal energy becomes too great for the substance to remain in a liquid state. When water is heated, it turns into a gas called water vapor. If the water is heated to the boiling point (212° F; 100° C) in a confined container, the water vapor becomes pressurized and is called steam. Because steam is under pressure, it can do work. Steam turbines spin under the force of flowing steam. A steam turbine works like a windmill, except the moving gas is pressurized water vapor instead of air. One of the first steam-powered devices was a turbine thought to have been built by the Greek philosopher Heron (also called Hero) of Alexandria in the first century A.D. It was used as a toy and served no practical purpose.

Our Findings

1. When the tea kettle got hot, a small white cloud should have started appearing at the spout. When it started whistling, the cloud began moving like a jet out of the top of the kettle.
2. When the bottle was placed in the bowl of hot water, the balloon inflated.
3. When the bottle was cooled, the balloon deflated.
4. The jet of steam coming out of the tea kettle could be used to turn a turbine. The expanding and contracting air in the bottle could be used to move a piston.

POWER BY PISTONS

It's not clear who first thought to use expanding and contracting gases as a power source, but the idea has been around for at least 2,000 years. In about 250 B.C., a number of Greek philosophers including Philon of Byzantium and Ctesibius of Alexandria experimented with changing the volume of air to make devices move. Ctesibius built an organ powered by compressed air. The pressure for the air came from a device called a piston.

Though the idea of a piston dates back to the ancient Greeks, the true value of pistons for making machinery run wasn't realized until the 1600s. A group of scientists and engineers living in Europe began using pistons to demonstrate the force of air pressure. In 1698, an English engineer named Thomas Savery designed the first practical piston-powered steam engine. It was used to pump water out of mines. Within a few short years, other engineers—including Thomas Newcomen and James Watt—improved on Savery's design. By the early 1800s, piston-powered steam engines were at work in mills, in factories, on steamships, and at railroads. They could be set up anywhere, so people no longer had to depend on water, wind, or animal power. The steam engine was the driving force behind the Industrial Revolution: It was all due to the power of the piston. In **Experiment 19: *Piston Motion Changes Pressure***, you will build a simple piston to see how its motion can affect the pressure of the air inside a cylinder.

Piston Motion Changes Pressure

Topic

How does the motion of a piston change the pressure of the gas in a cylinder?

Introduction

Pistons are important components in many modern-day machines. Pistons work by sliding up and down in a circular chamber called a cylinder. As a piston moves, it changes the volume of the gas that's trapped inside the cylinder. This changes the pressure of the gas, allowing it to do work. In this activity you will test to see how the volume and pressure of a gas change when the gas is trapped inside a cylinder with a moving piston.

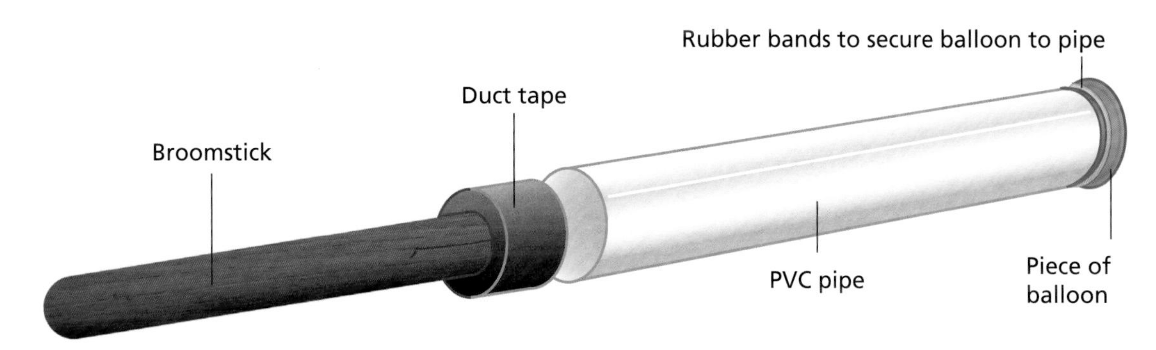

Figure 1

Time Required

30 minutes

Materials

- piece of hard PVC pipe about 1 ½ in. (4 cm) in diameter and 2 ft (60 cm) long (available at most plumbing, hardware or home improvement stores)
- wooden dowel, 2 ft (60 cm) long and about 1 in. (2.5 cm) in diameter (an old broomstick works well, but it must be narrower than the inside opening of the PVC pipe)
- round rubber balloon that inflates to between 9 inches and 12 inches (23 to 30 cm) in diameter
- four large, thick rubber bands
- roll of duct tape or black plastic electrical tape
- 12-in. (30 cm) piece of wax paper
- scissors

Safety Note Please review and follow the safety guidelines.

Procedure

1. Use the scissors to cut off the valve end of the balloon. You should be left with a large curved piece of rubber. Stretch the open end of the balloon so that it fits over one end of the PVC pipe. Wrap three of the rubber bands around it to secure it to the outside of the pipe.
2. Insert the wooden stick slowly into the open end of the pipe. It should easily slide in and out of the pipe. If the stick doesn't fit, find one that is narrower. Push the stick in about half way and observe what happens to the balloon.
3. Remove the stick from the pipe. Wrap one end of the stick with the tape so that it forms a knob. The diameter of the stick with the tape around it should be slightly smaller than the diameter of the inside of the pipe. While wrapping the tape, stop every few turns and check the thickness by inserting the taped end of the stick into the pipe. Stop when the taped knob almost completely fills the inner diameter of the pipe.
4. Insert the taped end of the stick into the pipe so that it is just inside the opening. Hold the pipe firmly in one hand and use the other hand to

quickly push the stick about halfway into the pipe. Observe what happens to the balloon when you push in the stick. Next, quickly pull the stick back out of the pipe. Stop when the taped end just reaches the opening to the pipe. Observe what happens to the balloon.

5. Remove the stick from the pipe. Use the wax paper to cover the taped end of the stick. Use the fourth rubber band to secure the waxed paper to the stick directly behind the taped knob. Insert the stick into the pipe and repeat Step 4. Note any differences in the behavior of the balloon and the stick as you move it in and out of the pipe.

Analysis

1. What happened to the balloon when you inserted the stick without the tape on it into the pipe?
2. What happened to the balloon when you inserted the taped end of the stick into the pipe? What happened when you pulled the stick out? Why?
3. How did putting the wax paper over the tape change the behavior of the balloon and stick when you moved it in and out of the pipe? Why?
4. Based on this experiment, why is it important for a piston to fit properly inside the cylinder?

What's Going On?

In this activity, the stick with the tape on it represented a piston. The rigid piece of pipe functioned as the cylinder. For a piston to work properly, its diameter must be only slightly less than the cylinder in which it rests. As a piston moves into a cylinder, the gas inside the cylinder is trapped. As the volume of the cylinder decreases, the pressure of the gas increases. The gas has become compressed. If the piston is too narrow for the cylinder, the gas will simply flow around the piston and out of the cylinder.

When you pushed the stick without the tape on it into the cylinder, there was little if any change in the balloon because the air was able to escape the cylinder around the piston. After taping the end of the stick, the piston was tight in the cylinder and the balloon inflated as you pushed the piston in. This is what happens in a device like a bicycle pump. You use mechanical energy to press down on the pump handle and the piston forces the air out the other end.

Pistons also can work in the opposite way. In a steam engine, steam under pressure is allowed to flow into a cylinder through an intake valve next to a piston. The steam pushes the piston up into the cylinder and a rod attached to the end of the piston turns a crank or wheel. Once the piston is at the far end of the cylinder, an exhaust valve opens. This lets the steam out of the cylinder and the piston returns to its original position. The exhaust valve then closes and another burst of steam comes in through the intact valve, repeating the cycle.

The action of the piston sliding back and forth in the cylinder generates a great deal of friction. This can cause the piston to heat up and expand. If the piston expands too much, it will no longer fit the cylinder, and the engine will stop. To reduce the chance of this happening, oil or some other lubricant is used in the cylinder. This decreases the amount of friction and helps the engine run cooler. In this activity, covering the end of the stick in wax paper served the same function as a lubricant.

Our Findings

1. Without the tape on the end, the balloon did not inflate when the stick was pushed into the pipe.
2. When the tape end of the stick was inserted into the pipe, the balloon inflated. When it was removed, the balloon was "sucked" into the pipe. This happened because the air could no longer move around the stick.
3. The wax paper made the stick slide in and out of the pipe easier. It acted like a lubricant to reduce friction. The wax paper also created a tighter seal between the piston and the cylinder, causing the balloon to inflate and deflate more quickly when the stick was moved in and out of the pipe.
4. If a piston doesn't fit the cylinder, the gas will be able to move around it, and it will be impossible to compress the gas.

PISTONS IN THE MODERN WORLD

Though steam engines are considered a technology of the past, there are plenty of places in our modern world where pistons are found. Bicycle pumps and balloon pumps rely on pistons to make them work, as do the compressors in refrigerators, freezers, and air conditioners. Without a doubt, the most common place pistons are at work today is in the engines of cars, trucks, motorcycles, and even lawn mowers. Most of these vehicles have internal combustion engines, which use pistons to make them work.

Internal combustion engines are smaller, lighter, and much more efficient than steam engines. German engineer Nikolaus Otto developed the first practical internal combustion engine in 1878. Instead of having a large external boiler to make steam to push a piston, an internal combustion engine uses burning fuel. A **valve** feeds a mixture of fuel (such as gasoline) and air into the cylinder. The fuel mix is ignited by a spark plug, causing it to explode inside the cylinder. The force of the explosion and the heat it releases make the air in the cylinder quickly expand. The expanding air pushes the piston up in the cylinder—this is called the power stroke. The bottom of the piston is attached by a rod to a crank. The crank pulls the piston back down in the cylinder. When this happens, the piston forces the exhaust gas out through a second valve and the cycle starts again.

Early piston-powered engines did have problems. First, because the power stroke of a piston happens in only one direction, engineers had to figure out a way to even out the power delivered by the engine between strokes. They also needed to keep the crank moving in order to get the piston to return to its original position in the cylinder. Both of these issues were solved with the introduction of a device called a flywheel. In **Experiment 20: *Managing Flywheel Momentum*,** you'll have the opportunity to experiment with a simple flywheel and discover how changes in the mass of the wheel affect its ability to keep an engine turning.

Managing Flywheel Momentum

Topic

How does the mass of a flywheel affect its ability to keep turning?

Introduction

Flywheels are important components in many modern-day machines. Generally made of heavy metal, such as iron or steel, a flywheel uses **momentum** to keep moving. All moving objects have some momentum. When you coast on a bike or glide on a skateboard, it is your momentum that is keeping you going. In both of these cases, your momentum is taking you in a straight path. With a flywheel, the momentum makes the wheel spin. This is known as angular momentum. The more angular momentum a spinning object has, the longer it will turn, and the more difficult it will be to make it stop. In this activity, you will test to see how mass affects the angular momentum of a flywheel by experimenting with a top.

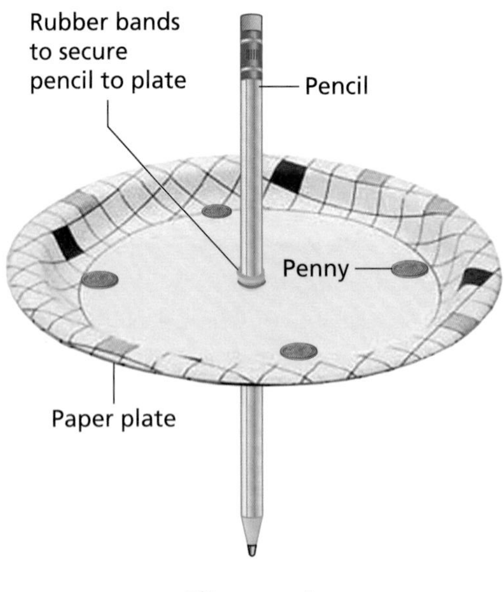

Figure 1

Time Required

30 minutes

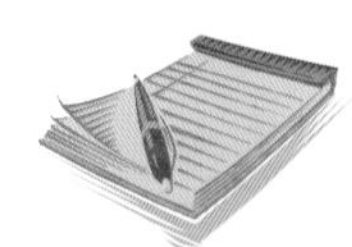

Materials

- heavy paper plate about 6 in. (15 cm) in diameter
- 2 large, thick rubber bands
- new, sharpened pencil
- pushpin or thumbtack
- ruler
- clock or watch with a second hand
- 8 pennies
- roll of cellophane tape

Safety Note **Please review and follow the safety guidelines.**

Procedure

1. Use the pencil and ruler to draw a line across the back of the plate that divides the plate in half. Draw a second line at a right angle to the first one, so that the plate is divided into quarters. The point at which the two lines intersect should be the middle of the plate. Using the point of the pushpin, carefully poke a hole through the plate at its center.
2. Turn the plate face up. Push the point of the pencil through the center hole so that about 2 in. (5 cm) of the pencil sticks out through the bottom of the plate. Be careful not to poke yourself with the point of the pencil. Wrap one rubber band around the pencil on each side of the plate. Push the two rubber bands so that they are tight against the plate. They should be tight enough to keep the plate from moving on the pencil.
3. Prepare to spin the top. Hold the pencil point down on a smooth, flat surface. Twist the pencil between your fingers and let go. The top should

begin spinning on its own. Test the launch procedure several times so that you feel comfortable. If you notice a big wobble in the top, check to make sure that the pencil goes through the center of the plate. If it doesn't, you will have to get another plate and try again.

4. You will test the top to see how long it spins under different conditions. Try to use the same amount of force to spin the top each time. In the first test, you will simply spin the top. Time how many seconds it spins and record it on the data sheet. Repeat the trial two more times to make sure you get consistent results.
5. Use the tape to secure four pennies to the top of the plate. The pennies should be opposite one another so that they form a plus sign (+, see Figure 1). Set each penny 2 in. (5 cm) from the center of the plate. After you have secured the pennies, spin the top again, using the same amount of force that you did in the first test. Do three trials and record your results on the data sheet.
6. Secure four more pennies to the top. They should be positioned between the first four pennies. When you are done, all eight pennies should be evenly spaced on the top. Using the same amount of force that you did in the first two tests, spin the top three times and record your results on the data sheet.
7. Remove four pennies from the right side of the top and leave four attached to the left side. Try spinning the top this time and note any differences from the first three tests.

Date Table 1			
	Trial #1	**Trial #2**	**Trial #3**
Top with no pennies			
Top with four pennies			
Top with eight pennies			

Analysis

1. Why is it important that the hole for the axle in the flywheel be exactly in the center of the wheel?
2. How did the number of pennies affect the length of time that the top spun?
3. What happened to the top when you had four pennies on only one side?
4. Based on your observations with the top, how do the amount and distribution of the mass of a flywheel affect the way it spins?

What's Going On?

In this activity, the spinning top performed like the flywheel in an engine. The greater the mass of a flywheel, the longer and steadier it will spin. This is because heavier objects have more momentum. Momentum is calculated by multiplying the mass of a moving object by its velocity. You can increase momentum of a moving object by increasing either its velocity or its mass. In early steam engines, it was discovered that increasing the velocity often had catastrophic consequences. Rapidly spinning flywheels often would break apart or explode, injuring people and wrecking factories. By slowing the wheel and making it heavier, engineers found that they could get the same momentum out of the wheel without as much potential for disaster.

When adding mass to a flywheel, it is important that it be evenly distributed on the wheel. If one side of the wheel is heavier than the other, the wheel will be out of balance. This has the same effect as an off-center axis. Not only will the wheel wobble, which wastes much of its stored energy, but an unbalanced flywheel also has a tendency to break apart.

Our Findings

1. If the pencil isn't exactly centered, the top will wobble and lose momentum more quickly.
2. In general, the more pennies that were on the top, the longer the top kept spinning.
3. When all the pennies were on one side, the top could barely spin at all. It just wobbled and fell over.
4. In a flywheel, the greater the mass, the more momentum it has and the longer it keeps spinning, provided the mass is evenly distributed.

PUTTING A CHARGE INTO MOTORS

So far in this chapter we've seen how the power used to drive machines has evolved over the years, from waterwheels and windmills to steam engines and turbines. For the most part, each innovation built on the technology that came before it. Animal capstans led to waterwheels, waterwheels led to turbines, and pistons came from the concept of expanding gases. In 1799, however, a power source was discovered that would take tools and machines in a whole new direction. That year, an Italian scientist named Alessandro Volta discovered the phenomenon known as electrical current. In the process, he invented the first battery.

Volta did not discover electricity. People had known about static electricity for thousands of years. By the mid-1700s, many famous scientists had experimented with static and a few even attempted to use it to power very simple machines. The problem with static electricity has to do with the way the electrical charge moves. All forms of electricity involve the motion of tiny particles called electrons.

Building on Alessandro Volta's discovery of electric current, Hans Christian Ørsted went on to discover that electric currents can create magnetic fields.

Electrons are part of atoms, so they are found in all types of matter. Static electricity happens when electrons start to build up on an object. When too many electrons build up in one location, they repel one another and move. Then, the object becomes "discharged." If you've ever gotten a shock after walking across a carpet, you've experienced static discharge. All the electrons move together in a single burst of energy. This makes it difficult to work with as a power source.

Electrical current also involves electrons in motion. In this case, instead of the electrons moving all at once, they move in a steady flow, like water moving in a stream. In building his first battery, Volta discovered that he could control the flow of electrons by using a chemical reaction involving metals and acid.

Once other scientists learned about Volta's discovery, they began experimenting with this new source of power. In 1820, a Danish scientist named Hans Christian Ørsted discovered that an electrical current flowing through a loop of wire would turn the wire into a magnet. This first **electromagnet** would become the basis of the electric motor, a device that revolutionized machines in the twentieth century. In **Experiment 21: *Making a Mini Motor***, you will discover how electromagnets are used to put the spin in electric motors.

Making a Mini Motor

Topic

How do electromagnets make an electric motor spin?

Introduction

Electric motors provide the driving force in many modern tools and machines. British scientist Michael Faraday built the first electric motor in 1821. It was little more than a spinning wire suspended above a bowl of liquid mercury. But it proved that current electricity could be used to put objects in motion. The key to making an electric motor spin lies in the direction of a magnetic field produced by an electromagnet. An electromagnet is only magnetic when an electric current is moving through it. The simplest electromagnet is a coil of wire wrapped around an iron or steel core. In this activity, you will use an electromagnet to show how it can keep an electric motor spinning.

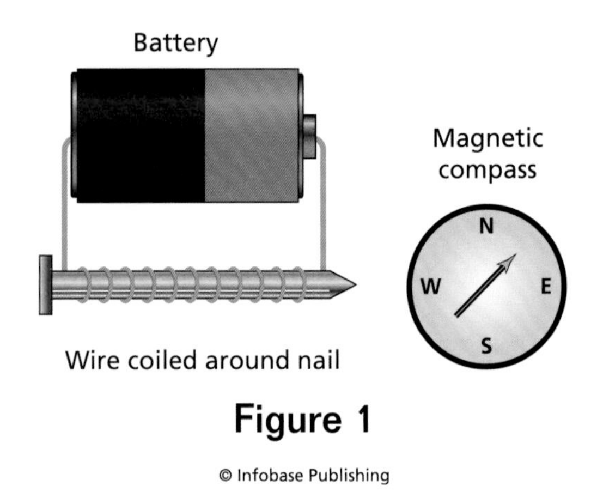

Figure 1

Time Required

30 minutes

Materials

- large steel nail about 3 in. (8 cm) long
- 24 in. (60 cm) piece of insulated wire with the two ends stripped
- AA, C, or D battery
- inexpensive magnetic compass
- 3 steel paper clips

Safety Note **When the two ends of the wire are touched to the battery, the wire may get hot. Be careful not to burn your fingers! Please review and follow the safety guidelines.**

Procedure

1. Wrap the wire around the nail to form a tight coil with at least 15 turns in it. The nail should look like the drawing in Figure 1 (without the battery). Leave about 4 in. (10 cm) of wire at each end of the coil free to allow you to attach the wire to the battery.
2. Without connecting the wire to the battery, touch the nail to the paper clips to see if the nail is magnetic. Touch the two ends of the wire to the two ends of the battery and bring the nail near the paper clips again. Observe what happens this time.
3. Lay the compass flat on a desk, table, or similar level surface. Without connecting the wires of the coil to the battery, lay the nail down next to the compass with the point touching the side of the compass. Observe what happens to the compass needle.
4. Without moving the nail, connect the battery to the two ends of the wire coil. Observe what happens to the compass needle when the electricity flows through the wire. (See Figure 1.)
5. Without moving the nail, disconnect the battery from the coil and turn it around. Reconnect the battery to the coil so that the top and bottom ends of the battery are opposite the way they were in Step 4. Observe what happens to the compass needle when the electricity flows through the wire this time.

6. Disconnect the battery from the coil and repeat Steps 4 and 5 several more times. Observe what happens to the needle each time you reverse the direction of the battery.

Analysis

1. What happened when you touched the nail to the paper clips when the wire was not connected to the battery?
2. What happened when you connected the battery to the coil and touched the paper clips?
3. What happened to the compass needle when you connected the coil to the battery in Step 4?
4. What happened to the compass needle when you reversed the direction of the battery and re-connected to the coil in Step 5?
5. Based on your observations, how might the compass needle in this experiment be similar to an electric motor?

What's Going On?

An electric motor spins due to the interaction of two magnetic fields. Every magnet, including an electromagnet, has two ends called poles. One end is the north pole and the other is the south pole. When two magnets are brought near each other, the opposite poles will attract and the same poles will repel or push apart. A compass needle is really a sensitive bar magnet that is free to turn in a circle. When the electric current ran through the coil surrounding the nail, it turned the nail into a temporary electromagnet, and one end of the compass needle was attracted to the point of the nail. When the battery was turned around, the current flowing through the wire coil reversed direction. This reversed the poles of the electromagnet. As a result, the opposite end of the compass needle was then attracted to the point of the nail.

An electric motor works on the same principle. The most basic electric motor has two types of magnets. On the outside of a central hub are fixed, permanent magnets with their opposite poles directed inward. Between the permanent magnets is an electromagnet connected to a rotating shaft. When current is sent through the electromagnet, its poles interact with the poles of the fixed permanent magnets, and the shaft turns. The trick to keeping the inner shaft spinning is reversing the direction of the current in the electromagnet every half turn. This is done by a device called a commutator, which

American inventor Joseph Henry developed in 1835. Within a few short years, electric motors were being put to use in factories and workshops all over the world.

Our Findings

1. When the coil of wire was not connected to the battery, the nail was not magnetized and it did not attract any paper clips.
2. When the battery was connected to the coil, the nail became an electromagnet.
3. When the electromagnet was first brought near the compass, one end of the compass needle became attracted to the point of the nail.
4. When the direction of the battery was reversed, the opposite end of the compass needle was attracted to the point of the nail.
5. The compass needle is like the shaft of a motor, spinning under the influence of two different magnets.

ELECTRIC MOTORS IN THE MODERN WORLD

Electric motors can be found at work in almost every room of the house. Blenders, dishwashers, washing machines, and clothes dryers all have electric motors in them. This is also true of powered toothbrushes, electric shavers, and blow dryers. Electric motors can be found in DVD players and camcorders, as well as computer printers and copy machines. In the workshop, many power tools are run by electric motors, including drills, saws, routers, and sanders. Electric lawn mowers and hedge clippers have them, and so do weed cutters. So many devices are powered with electric motors that it's hard not to find one somewhere close by.

FROM ELECTRIC MOTOR TO ELECTRIC GENERATOR

Today, our lives are filled with devices that run on current electricity. Where does all the electricity come from? Back in the early 1800s, the only way that people could produce electrical current was by using the chemical reactions in batteries. That worked fine for small devices, but large motors required enormous batteries, which eventually run out of power. People needed another way to produce large quantities of electrical current without having to depend on batteries. In 1831, both Joseph Henry and Michael Faraday solved the problem.

Henry and Faraday independently discovered that if a magnet is moved past a coil of wire, the magnet will "induce" an electric current in the wire. Because an electric motor is really a coil of wire surround by magnets, spinning the shaft of a motor without attaching it to a source of electricity would cause the motor to make electricity. In other words, a motor turned backwards is really an electric generator. All that was needed was a way to turn the shaft. This could be done with a hand crank, a windmill, a waterwheel, or even a steam engine. As mentioned earlier in this chapter, most modern power plants use steam turbines to turn their generators. Because of problems with global warming and the rising cost of fossil fuels such as oil and gas, a growing number of utilities are going back to "natural sources" of energy such as wind and water to keep their generators spinning.

5

Staying in Control

As tools and machines became more complex, another problem arose. People had to keep these new devices under control. Control was not really an issue with simple hand tools. If a person who was using a tool wanted it to speed up, slow down, or stop, he or she simply sped up, slowed down, or stopped. The user was in direct control. With animal power-machines, things were a little more complicated. The operator had to convince the animal to speed up, slow down, or stop. This was usually accomplished with a tug on a harness, a crack of a whip, or a treat for the animal to eat.

After waterwheels and windmills were put into use, the task of controlling machines became more difficult. People had to monitor the machinery to make sure it ran properly. If a machine turned too fast or too slow, the operator had to physically adjust the speed by changing gears, applying a brake, or redirecting the flow of water or wind. With the introduction of the steam engine, control of machinery became even more critical. Workers had to regularly check the water level in the boiler to make sure it didn't run too low. They also had to watch the steam pressure, because if it got too high, the boiler could blow up. In many cases, the lack of control of a steam engine resulted in entire buildings being leveled and lives being lost.

To help reduce control problems, inventors have come up with devices that allow people to monitor and measure the performance of machines. Most of these devices are either **gauges** or **meters**. They are found in many of the devices we use every day. In **Experiment 22: *Gauging the Pressure*,** you will test how a simple air pressure gauge helps you judge if the tires on your bike are in working order. In **Experiment 23: *Making a Meter*,** you will discover how meters use the force of electromagnets to measure the flow of current.

Gauging the Pressure

Topic

How do gauges allow you to more accurately judge the performance of a device?

Introduction

Gauges and meters are designed to give a machine operator a better understanding of how the machine, or a certain part of the machine, is functioning. Gauges are usually designed to present information on some type of scale that compares the actual condition to some known standard. Often, a machine will have a number of gauges at work. Automobiles usually have an engine temperature gauge, an oil pressure gauge, and a fuel gauge. In this activity, you will discover why gauges are important by testing your own ability to judge the operational condition of a bicycle tire.

Time Required

30 minutes

Materials

- hand pump
- bicycle tire
- tire pressure gauge

Safety Note **Be careful not to over- inflate the tire. Please review and follow the safety guidelines.**

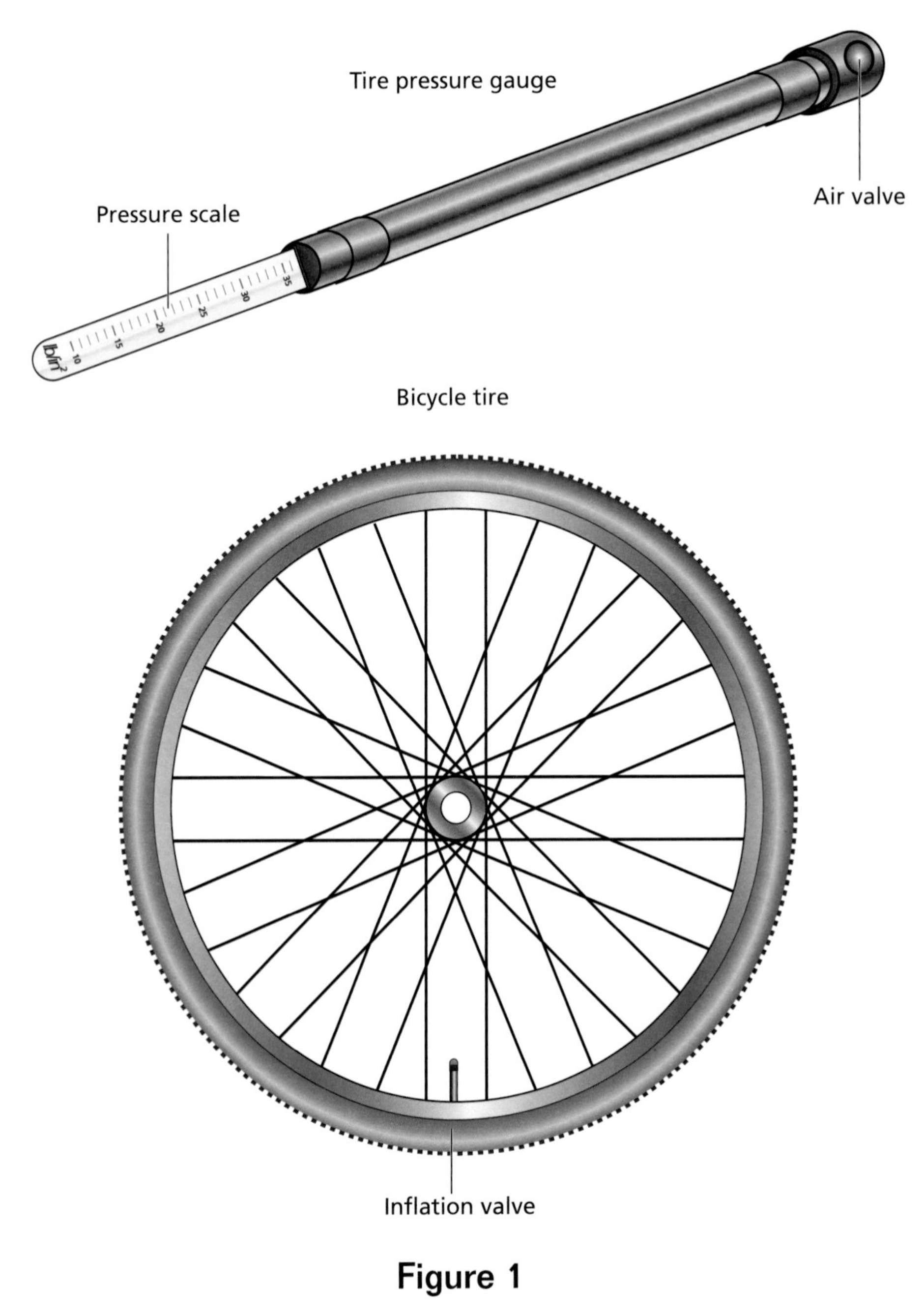

Figure 1

Procedure

1. Closely examine your bicycle tire to determine how many pounds of air pressure are needed to fully inflate it. This figure is usually printed on the sidewall of the tire, followed by the letters *PSI*. By looking at and feeling the tire, estimate how full you think the tire is.
2. Unscrew the cap on the tire inflation valve and place the tire pressure gauge over the valve. As the air comes out of the tire, the indicator on the

gauge will show how much air actually is in the tire. Observe how this number compares with the amount that should be in a fully inflated tire.

3. Using your finger or the little nub found on the back of the tire pressure gauge, let all of the air out of the tire. Check to see that the tire has no air pressure by using the tire pressure gauge on the valve again.
4. Place the valve of the pump on the tire valve and begin adding air to the tire. Stop when you think that the tire is half full. You can estimate this by looking at the tire and feeling it with your hands. Then, use the air pressure gauge to check the pressure. If your estimate was accurate, the pressure on the gauge should read half the pressure value for the fully inflated tire.
5. Continue inflating the tire until you think that it is completely full. As you did in Step 4, use the tire pressure gauge to see if your estimate was accurate. When you have finished, use the gauge and the pump to adjust the tire pressure to its proper amount and put the screw cap back on the tire valve.

Analysis

1. When you first measured the tire pressure, how close was your estimate to the measured value?
2. In Step 4, how close was your estimate of when the tire was half full to the actual value as measured by the tire pressure gauge?
3. In Step 5, how close did you come to estimating when the tire was fully inflated?
4. Based on this activity, why are gauges important for operators of machinery?

What's Going On?

Gauges are important because they present a quantitative, or numerical, measurement of a certain condition in a machine or device. They usually are more accurate than a qualitative description of the same condition. A gauge takes the guesswork out of a measurement. This is particularly important with certain machines designed to operate under a narrow range of conditions. For example, many machines use steam as a source of power. Steam is usually made in a device called a boiler. All boilers have a gauge on them so that the operator can know how much pressure the steam has. If the steam pressure in a boiler gets too great, the boiler could explode. If, on the other hand, the steam pressure is too low, it will not have enough energy to power the machinery.

Our Findings

1. The answer will vary, but in most cases the estimated value will not exactly match the measured value.
2. The answer will depend on how good you are at estimating tire pressure.
3. Your answer was probably closer to the true value than it was in Step 1 because you had some information to base it on from the previous step.
4. Gauges give machine operators readings of the different conditions under which a machine is running.

Making a Meter

Topic

How do meters work to monitor the flow of electric current?

Introduction

Meters are similar to gauges in that they are both designed to give the operator a better understanding of how a machine or a certain part of the machine is functioning. Most meters have either a digital or analog (dial) readout that displays a series of numbers. A car, for instance, has a speedometer, which shows how fast the car is moving, and an odometer, which shows how far the car has traveled. Many cars also have a tachometer, which measures how fast the engine is turning, and an ammeter, which tells how much electrical power is flowing out of a battery. Most meters run on electric current. In this activity, you will build and test a simple electric current meter using a compass, coil of wire, and several batteries.

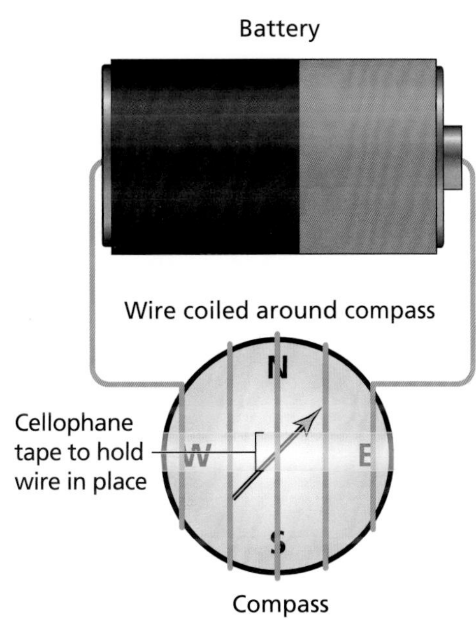

Figure 1

Time Required

30 minutes

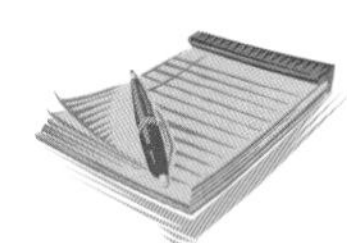

Materials

- new AA, C, or D battery
- used (weak) AA, C, or D battery
- dead AA, C, or D battery
- ruler
- inexpensive magnetic compass
- cellophane tape
- 24-in.-long (60 cm) insulated bell or wire hook-up with ends stripped

Safety Note **When the wire is connected to the two ends of the battery, it will get warm. Be careful not to burn your fingers. Please review and follow the safety guidelines.**

Procedure

1. Measure 6 in. (15 cm) from the end of the wire. Starting here, make five parallel loops of wire around the face of the compass. Each loop should touch the one next to it, but the wires should not cross. After making the loops, you should be left with about 6 in. (15 cm) of additional wire. Use the tape to secure the wire to the compass.
2. Lay the compass flat on a table. Turn it so that the compass needle is parallel to the wire loops above it.
3. Without moving the compass, attach one end of the wire to the (+) end of the new battery and the other end of the wire to the (–) end. Observe and record what happens to the compass needle when the battery is connected to the wire.

4. Disconnect the new battery and repeat Step 3 using the weak battery, and then the dead battery. Observe and record what happens to the needle each time you change the battery.

Analysis

1. What happened to the compass needle when you connected the fresh battery to the wire coil?
2. How did the compass needle react when you connected the weak battery to the wire coil?
3. How did the compass needle react when you connected the dead battery to the wire coil?
4. Based on your observations, what relationship exists between the movement of the compass needle and the strength of the battery?

What's Going On?

Many power tools and machines have meters that allow the operator to monitor performance. These meters may serve many functions, but they usually base their measurements on the amount of electricity flowing through a circuit. The stronger the electrical current flowing through the circuit, the higher the reading on the meter. The simplest form of current meter is called a galvanometer.

A galvanometer works because of the interaction of an electromagnet and a permanent magnet. When electricity flows through a coil or loop of wire, it creates a magnetic field around the wire. This is called electromagnetism, and it happens only while the electrical current is flowing through the wire. The stronger the electric current is, the stronger the magnetic field surrounding the coil will be. An electromagnetic field has two ends called poles. When two magnets are brought near each other, their opposite poles will attract and their similar poles will repel. When an electric current flows through a galvanometer, the magnetic field of the wire coil and the magnetic field of the permanent magnet interact with each other, causing either the permanent magnet or the coil to move. This motion is transferred to a needle on the meter. The stronger the current flowing through the galvanometer, the greater the interaction between the two magnets, and the more the needle moves.

Our Findings

1. When the coil was connected to the new battery, the compass needle should have turned so that it was perpendicular (at right angles) to the loops of wire.
2. With the weak battery, the needle of the compass should have moved, but not as far as it did when connected to the new battery.
3. When the dead battery was connected, the needle of the compass should have remained parallel to the loops of wire.
4. The stronger the current flowing through the wire loop (coil), the more the needle of the compass will move.

CONTROLLING THE FLOW

Gauges and meters let an operator monitor the performance of a machine, but that's only part of the control process. Often, the person running the machine has to take some type of action to keep the machine running properly. In many machines, the main control device is a valve. Valves come in hundreds of shapes and sizes, but all are designed to control the flow of a liquid or gas.

In **Experiment 22: *Gauging the Pressure*,** you used two valves to control the flow of air. One valve let the air in and out of the tire. The second valve let the air in and out of the tire pressure gauge. Valves also are used to control the flow of liquids. To make a car go faster, the driver steps on the gas pedal. This either directly or indirectly opens a valve, allowing more fuel to enter the engine. Valves are a big part of our daily lives. Without them, you would have a hard time washing your hands or using a toilet. Every time you turn on a faucet or flush a toilet, you are using valves. In **Experiment 24: *How Valves Control Liquid Flow*,** you will discover how one type of valve controls the flow of a liquid.

EXPERIMENT 24 How Valves Control Liquid Flow

Topic

How do valves work to control the flow of a liquid?

Introduction

Valves are important control devices found in many types of tools and machines. The main function of a valve is to control the flow of steam, gasoline, water, or air. Most valves work by restricting the flow of the gas or liquid in a closed pipe or chamber. This is done in a variety of ways. In this activity, you are going to construct a simple valve and see how well it controls the flow of water between two containers.

Time Required

45 minutes

Materials

- 1-liter soda bottle
- new, sharpened pencil
- roll of duct tape
- scissors
- 2 16-oz wide-mouth plastic cups
- water
- ruler
- sink
- pencil

Safety Note Please review and follow the safety guidelines.

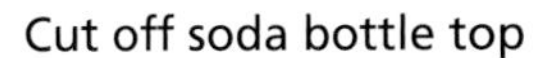

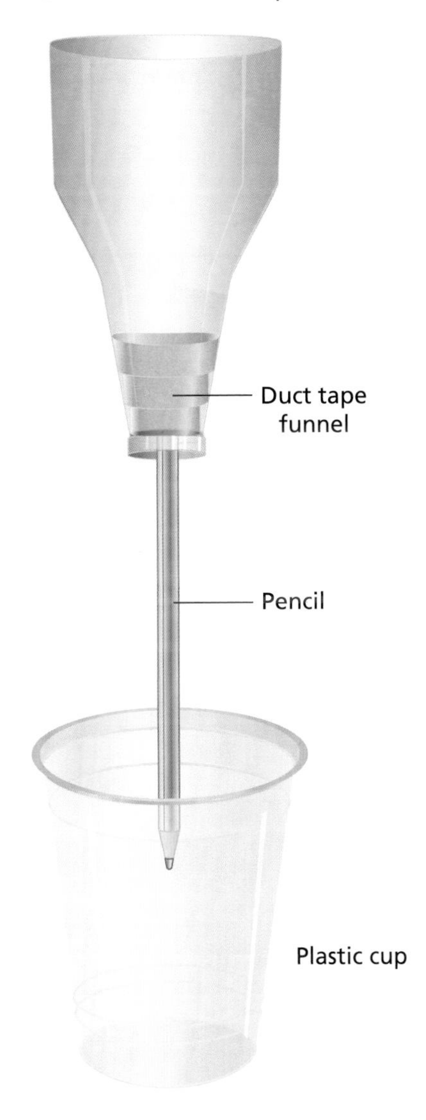

Figure 1

Procedure

1. Use the scissors to remove the label from the soda bottle. Measure 8 in. (20 cm) down from the top of the bottle and mark the spot with the pencil (or a pen). Cut the bottom off the bottle at the pencil mark. The top of the bottle should look like a funnel.
2. Cut five pieces of duct tape, each 6 in. (15 cm) long. Place them along the edge of a counter or table so they don't get folded or stuck together. Wrap the first piece of tape around the pencil so that one edge of the tape is even with the eraser at the top of the pencil. Wrap a second piece of tape around the pencil. The bottom of this piece should be $4/5$ or $8/10$ in. (2 cm) higher than the bottom of the first piece. The top edge of this piece should

overlap the top of the pencil by about 4/5 or 8/10. Wrap a third piece of tape around the pencil 4/5 or 8/10 in. higher than the second piece. Continue this procedure until you have used all five pieces of tape. Use the scissors to trim the tape that overlaps the top of the pencil. The tape on the pencil should be in the shape of a cone, with the thinnest end near the pencil point and the widest near the eraser end (see Figure 1).

3. Hold the cut-off bottle top with the neck pointing down. Place the taped pencil, point-end down, inside the bottle. Allow the pencil to fall into the bottle. The taped end of the pencil should get stuck in the neck of the bottle like a stopper. If the pencil slides through the opening, wrap another piece of tape around the top to make it thicker. Then try again. Keep adding tape until the pencil does not pass through the opening.
4. Pull the pencil tightly down into the neck of the bottle so that it seals the opening. Hold the neck of the bottle over an empty plastic cup. Fill the second cup with water. Pour the water into the open end of the bottle. Observe what happens to the water in the bottle.
5. Lower the bottle top into the empty cup so that the point of the pencil just touches the bottom. Slowly push the bottle down so that the pencil slides up into the bottle about 1/4 in. (1/2 cm). Observe what happens to the water in the bottle. Lift the bottle back up so that the point of the pencil is no longer touching the bottom of the cup. Observe what happens to the water.
6. Empty the cup and replace the water in the bottle. Repeat Step 5, but this time, push the pencil down hard so that it moves up into the bottle about 4/5 or 8/10 in. (2 cm). Compare how water flows this time with the way it flowed in Step 5.
7. Empty the water from the cup and the bottle into the sink. Turn on the cold water faucet just a trickle. Observe the flow of the water as you gradually turn the faucet until it is open all the way. Then turn off the faucet.

Analysis

1. What happened to the water in the bottle when the pencil was stuck tight in the neck of the bottle?
2. What happened to the water when you pushed the pencil up slightly into the bottle? What happened to the pencil when you lifted the bottle back up?
3. What happened to the water in the bottle when you pushed the pencil far into the bottle? How did this compare with the flow in Step 5? Why?

4. Based on your observations in this activity, what do you think is happening inside a faucet when you turn the water on and off?

What's Going On?

In this activity, you built a simple model of a device called a poppet valve. Valves control the flow of a liquid or gas by restricting an opening through which it is designed to pass. In poppet valves, a central core called the valve stem acts like a stopper in the neck of a bottle. When the valve stem is pushed up, it creates an opening through which the liquid or gas can flow. When the pressure on the valve stem is released, it moves back to its original position. This blocks the opening and stops the flow.

In this activity, the taped pencil acts like a valve stem. In poppet valves, the top of the valve stem is tapered; as it is pushed higher, the opening enlarges, and more liquid or gas can flow through. Poppet valves are commonly found in car engines, where they control the flow of gases in and out of the cylinders.

Our Findings

1. When the pencil was stuck tight in the bottle, no water came out.
2. When the pencil moved up into the bottle, a small amount of water came out. When the bottle was lifted, the pencil was forced back into the neck of the bottle, stopping the flow of water.
3. When the pencil was pushed higher into the bottle, the flow of water increased because the opening got bigger.
4. When you turn on a faucet, a device like a stopper moves out of the valve. This creates an opening through which water can flow. When you turn the faucet off, the stopper moves back into place.

VALVES IN THE MODERN WORLD

Valves can be found at work in just about any device that involves the flow of a liquid or gas. To accomplish a range of tasks, engineers have created valves with different designs. The faucets in your house and the control valves on steam radiators are called globe valves. The valve stem usually has a tapered plug or disk on the end, which rests against an opening in the valve called a seat. When the faucet is turned on, a screw on the valve stem pulls the plug off the seat, opening the valve and allowing the fluid to flow. When the screw is turned in the opposite direction, the valve stem is forced down, closing the valve.

Butterfly valves don't have a valve stem. Instead, the control element is a circular disk called a butterfly, which has a rod running through its center. The disk is set inside a circular chamber at an angle. It is designed to "flap" back and forth, opening and closing the valve. These types of valves usually control the flow of air. They are commonly found in stovepipes, heating vents, and the carburetors of internal combustion engines.

In many machines, it is important that the liquid or gas flows in only one direction. To solve this problem, engineers have come up with check valves. In most check valves, the control element looks like an angled flap that is controlled by the pressure of a liquid or gas behind it. If for some reason the pressure drops, the valve automatically closes. This keeps the liquid or gas from backing up in the pipe. Check valves have a range of uses. They are commonly found in the compressors of refrigerators and air conditioners, in bicycle pumps, and even in toilet bowls.

When the steam engine was first put into use, one of the most important improvements was the safety valve. Safety valves were placed on boilers to keep them from accidentally exploding if the steam pressure got too high. Most safety valves are designed to work like the poppet valve you built in **Experiment 24: *How Valves Control Liquid Flow*.** They have a valve stem held in place by a spring or weight. Most of the time, the valve is closed. If the pressure beneath the valve stem gets too high, the valve is forced open, releasing some of the steam. Though we don't have many steam engines in our homes, safety valves are still at work on radiators, pressure cookers, and even tea kettles.

CIRCUITS AND SWITCHES

Pipes and vents aren't the only places where it's important to control the flow. In electrically powered machines, the flow of the electrical current must be controlled, or the machine won't work properly. In many ways, the flow of electricity through a wire is similar to the flow of water through a pipe. In fact, when Thomas Edison was inventing many of the electrical devices we use today, he often would make a model using water. The models helped him understand the direction in which the current had to flow.

The basic design element of electrical devices is called a **circuit**. A circuit is a closed loop or pathway through which an electrical current can flow. Most circuits are made from wires, but they can include other components that are also **conductors**. A conductor is a material that allows electricity to pass through it. Metal is a conductor. **Insulators** are materials that stop the flow of electricity. Insulation (like the plastic coating on a wire) is used to keep electricity from flowing outside of a circuit. Some typical insulating materials are glass, plastic, rubber, and wood.

For an electric machine to run, the circuit must be complete. The electrical current has to return to the place where it entered the circuit. If there are any breaks or interruptions, the electricity stops flowing, and the machine stops. This is what a **switch** is designed to do. A switch intentionally breaks a circuit, allowing the operator to start and stop the machine.

Switches aren't the only control devices found on electric machines. In many cases, the operator needs to control the speed of the motor. This is done using a variable **resistor**. A resistor reduces, or resists, the flow of electricity in a circuit. In **Experiment 25: *Resistance Against Electricity Flow***, you will build a simple variable resistor and discover how it changes the amount of electrical current flowing through a circuit.

EXPERIMENT 25 Resistance Against Electricity Flow

Topic

How do variable resistors work to change the flow of electricity in a circuit?

Introduction

When people use electrically powered devices, they often have to control the current that flows through the circuit. One way of doing this is to control the resistance of different parts of the circuit. Resistance is a measure of how difficult it is for electricity to pass through a conductor. The higher the resistance, the harder it is for electricity to flow. One way to control the resistance in a wire is to make it narrower. Just as a narrow pipe restricts the amount of water that can flow through it, a narrow wire offers greater resistance to the flow of electricity.

Sometimes, such as when people dim a light or change the speed of an electric motor, there is a need to temporarily change the amount of electricity flowing in a circuit. This is done by adding a device called a variable resistor. A variable resistor acts like an electric valve. It increases and decreases the amount of electricity flowing through a circuit, just like a valve controls the

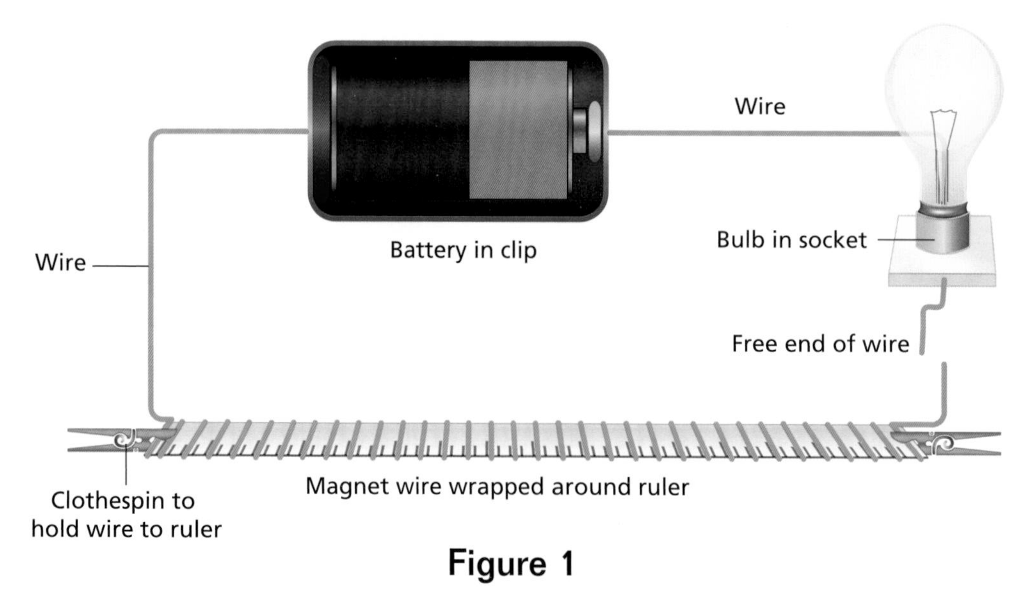

Figure 1

amount of water flowing through a pipe. In this activity, you will build and test a simple resistor to determine how it controls the flow of electricity.

Time Required

45 minutes

Materials

- 12-in.-long (30 cm) plastic ruler
- 50-ft (15 m) roll of 30-gauge enameled magnet wire
- piece of fine sandpaper
- scissors
- 2 clothespins
- D battery with holder
- flashlight bulb with socket
- 3 12-in. (30-cm) pieces of insulated bell or hook-up wire
- roll of cellophane tape

Safety Note **Please review and follow the safety guidelines.**

Procedure:

1. First you will make the resistor. Measure 2 in. (5 cm) of magnet wire from the end of the spool. Starting at this point, begin wrapping wire around the ruler so that you make a series of wire loops along its length. The loops should be spaced about $\frac{1}{12}$ in. (2 mm) apart, but they should not touch or overlap one another (see Figure 1). You will need at least 100 loops of wire. When you complete the loops, spool out another 2 in. (5 cm) of wire. Then, cut the wire off the spool with the scissors. Clip one clothespin over each end of the ruler to keep the wire from unraveling.

2. Flip the ruler over and run a piece of cellophane tape along its entire length to secure the wire to the ruler. Turn the ruler over to the front side. Use the sandpaper to gently rub the top of each wire loop to remove the insulating enamel. Do not rub too hard. If you do, you may break the wire. After you have removed the insulation from the tops of the wire loops, sand the two ends of the wire as well. When you are done sanding, put the resistor off to the side and build the rest of the circuit.
3. Use the scissors to strip ½ in. (1 cm) of insulation off each end of the three pieces of bell wire. Make certain that the battery is tight in the battery holder and the bulb is tight in the socket. Use one piece of wire to connect one side of the socket to the battery clip. Attach the second piece of wire to the other end of the socket and the third piece of wire to the other end of the battery holder. Test the circuit by touching the two loose ends of wire together. The bulb should light. If it does not, make certain that all the connections are tight and that only bare metal wire is in the connections (see Figure 1).
4. Connect the loose end of the wire coming from the socket to one end of the resistor by twisting the ends of the two wires together. Make certain that at the point of contact, all the enamel has been removed from the resistor wire. Take the loose end of the wire attached to the battery holder and touch it to the connection that you just made on the resistor. The bulb should light. Observe the brightness of the bulb and compare it to the brightness of the bulb in Step 3.
5. With one end of the resistor still connected to the socket, slowly move the end of the loose wire down the tops of the wire loops along the length of the ruler. Make contact with the magnet wire only in the places where you sanded off the insulating enamel. As you slide the wire down the length of the resistor, observe the brightness of the bulb.
6. When you reach the far end of the ruler, touch the loose end of the wire to the tip of the wire at the far end of the ruler. Observe the brightness of the bulb. Predict what will happen to the bulb brightness if you slide the wire back to the other end of the resistor. After you have made your prediction, move the wire back to the other end of the ruler to observe the results.

Analysis

1. How did the brightness of the bulb in Step 4 compare with the brightness of the bulb in Step 3?
2. What happened to the brightness of the bulb as you slid the loose end of the wire down the length of the ruler? Why do you think this happened?
3. When was the bulb the dimmest? Why?

4. What happened to the bulb brightness when you moved the wire back to the starting point on the ruler? Why?
5. Based on your observations in this activity, what do you think is happening inside a light dimmer or the speed control on an electric fan?

What's Going On?

In this activity, you built a device called a rheostat or potentiometer. Rheostats are important components in many electrically powered devices. In machines, they control the speed of the motor. In radios and televisions, they control the volume. In lights, they serve as dimmers. Rheostats control the amount of electricity flowing through a circuit by varying the resistance. Resistance in a wire is controlled by a number of factors, including the length of the wire, its thickness, and the metal from which it's made. In this activity, you changed the resistance by changing the length of the wire. As you slid the loose end of the wire from one end of the ruler to the other, the electricity had to flow through more wire loops. This increased the length of the wire through which the current flowed, which increased the resistance in the circuit. The result was a reduced flow of electrical current going to the bulb. As a result, the bulb got dimmer.

Instead of being straight like the ruler, many variable resistors in electric devices are circular. A rotating knob controls this type of resistor. The wire loops are wrapped around a curved, insulating fiber strip. A metal slider is connected to the knob. As it moves over the wire loops, it changes the resistance. In some rheostats, instead of using wire loops, the slider passes over a carbon rod to change resistance.

Our Findings

1. In Steps 3 and 4, the bulb should have the same brightness.
2. As you move the wire down the length of the ruler, the bulb should have dimmed. The electricity had to flow through more wire.
3. The bulb was the dimmest when the wire was attached to the opposite end of the ruler from where the socket was attached. The current had to flow through the most wire, which made the greatest resistance.
4. Moving the wire next to the first connection on the ruler made the bulb brighter because the wire got shorter, lowering the resistance.
5. As you turn the knob on a dimmer, you increase the resistance in the circuit, dimming the light.

SENSING THE SURROUNDINGS

Valves, switches, and variable resistors provide much of the control in modern machines, but they don't usually work alone. In many cases, these control devices are triggered by another component called a **sensor**. As the name suggests, a sensor works just like one of your senses. It gathers information about one or more operating conditions. It then compares this information against some pre-selected standard. Based on the comparison, some action is taken. Sensors gather information on a range of conditions, including wind speed, air pressure, steam pressure, salinity, and the amount of oxygen in the air. Two of the most common sensors used in machines today are thermostats and photoelectric cells.

Most people are familiar with thermostats because they are in most homes. Thermostats are designed to monitor temperature. They usually are connected to heating systems and air conditioners and designed to trigger a switch that turns the unit on and off if the temperature gets too hot or cold. Thermostats aren't just found in buildings. They frequently are found in engines and motors, where they keep track of the operating temperature. If the temperature gets too hot, the thermostat may trigger a switch that turns on an automatic cooling system. Or, it may shut down the device until it has cooled on its own.

Without a thermostat to regulate temperature, motors would burn up. Photoelectric cells sometimes are called "electric eyes." Just like human eyes, they sense light. When light strikes a photoelectric cell, it generates an electric current. The brighter the light, the greater the amount of electrical current produced. There is a great deal of interest in using photoelectric (or photovoltaic) cells to generate electricity from the sun. If you have a solar-powered device, its power comes from the sun. Photoelectric cells also can be used to trigger switches as light conditions change. Lights that automatically turn on when it gets dark use photoelectric cells in this way. They also are found in copiers, fax machines, and color printing presses. Photoelectric cells also are critical components of many smoke detectors.

Fabricating the Future

We've covered a great deal of ground. You have learned about the history and evolution of some of our most important tools and machines, as well as the science behind how they work. New machines seem to be popping up all around, doing many types of jobs. There are mammoth digging machines that can literally move mountains, and machines so tiny they can fit inside the human body. We've put machines up in space, under the oceans, and even inside active volcanoes. There even are machines that make other machines!

As we look into the future, it's difficult to know which way the evolution of tools and machines might be heading. Technological trends over the past two centuries offer some clues. Now let's look at some recent technological developments and where they may lead us in the future.

THE ACT OF AUTOMATION

Automation and mechanization occur when machines are used instead of people or animals to do jobs. These two terms are sometimes used interchangeably, but they are not the same. Mechanization occurs when a machine does a job instead of a person or animal. Under this definition, using almost any tool or machine would fit the bill. Mechanization has been at work for thousands of years, starting with first uses of gears, pulleys, and levers. It continues today.

Automation is a bit more complicated. In a truly automated machine, three things have to happen. First, the machine needs an outside power source that allows it to perform some function. Second, the machine needs a **feedback** system so that it can operate under its own control. Third, the machine needs to be **programmable**.

Most of the automated machines found in factories and businesses today are designed to use electricity as an energy source. This is because electricity can be used in so many different ways. It can be used to make heat, weld metal, and run motors that can, in turn, power pumps, belts, fans, and gears. Electricity also offers many advantages when it comes to con-

An assembly line involves automated machines moving items from one place to another, as shown in this automobile plant assembly line for Toyota Peugeot Citroen Automobile. The photo was taken in the Czech Republic in 2008.

trolling machines, as many meters, gauges, and sensors run on electricity.

Automated machines are used for hundreds of jobs. The two most common ones are processing materials and transferring materials from one place to another. Material processing covers a range of jobs. Giant mechanical presses bend large pieces of steel into bumpers for cars. Other machines pluck the feathers off chickens. Automated machines help to make almost any mass-produced item in use today.

As the name suggests, material transfer involves using automated machines to move items from one place to another. This is particularly important on assembly lines, where pieces are added to a product in a series of steps. Conveyor belts, chains of buckets, and specialized moveable arms are just some of the automated devices that are used to keep materials moving.

FEEDBACK

People usually think of "feedback" as the ear-splitting noise that happens when someone accidentally holds a microphone in front of a speaker. Though the word may be the same, the concept of feedback in automation is not about sound. A feedback system, or loop, is how a machine regulates itself.

Most feedback control systems have at least four parts. First, there is some sort of input. This is a value or values that act as a standard. It could be an ideal temperature, an optimum speed, or a maximum pressure. The next part involves sensors, which monitor the performance of the machine and gather data. This data is called the output, and the sensors send it on to a central controller. In the third part, the controller takes the output data coming from the sensors and compares it with the input values. If the output matches the input, no action is needed. If the output is different from the input, then the controller sends a signal to another device called an actuator. The actuator in some way changes the performance of the machine. At this point, the cycle starts again, with the sensors taking a new set of readings. It's called a feedback loop because the information from the controller is fed back to the machine to keep it under control.

In many modern machines, computers carry out feedback control. But long before there were computers, there were feedback systems. One of the first systems was used in the steam engine.

The operation of early steam engines created many problems. If the steam pressure got too high in the boiler or if the flywheel turned too fast, the engine could explode. Someone

The watt flyball governor on this steam pumping engine controls the pressure and speed of the engine.

had to watch the engine at all times, and adjust its pressure and speed by turning a valve. If the worker became distracted or left his post, there could be dire consequences. People needed a way to automate the control process.

In the 1780s, Scottish engineer James Watt found that he could use the speed of the engine to control the pressure of the engine. He designed a rotating collar for the steam valve. A belt connected the collar to the engine's flywheel. When the engine turned, so did the collar. The top of the collar had two heavy metal balls attached by levers. As the engine turned faster, the balls flew out farther. As they flew, the levers opened the valve, letting excess steam out of the boiler. This reduced the pressure and slowed the engine. At that point, the balls (called flyballs) slowed as well, which closed the valve. Soon, pressure began building and the engine began to speed up again.

As long as the engine was turning, Watt's invention—called the flyball governor—could control the pressure and speed. After the flyball governor was perfected, the steam engine became much more user-friendly. The invention not only saved lives and reduced property damage, but also helped to kick the Industrial Revolution into high gear.

All of the control devices discussed earlier are commonly used as feedback elements in modern machines. Gauges and meters act as sensors, providing data for electronic controllers. Thermostats and photoelectric cells are used to control switches when heat and light conditions change. Although the development of feedback systems helped to automate many machines throughout the ninteenth and twentieth centuries, the ability to program a machine really made automation work.

MACHINE PROGRAMMING

These days, people usually associate the word *program* with a computer. Yet, in an automated machine, a program is simply a set of commands that are followed in a certain order. Machine programming often is tied into a feedback system, because each new command is a new source of input that has to be evaluated and controlled. French inventor Joseph Marie Jacquard developed one of the first truly programmable machines in 1805. The Jacquard loom was used for making cloth with very complicated patterns. Before Jacquard, making cloth with complex patterns was extremely time consuming. If an operator wanted to change the pattern on a manual loom, she had to physically stop the loom and select which colored threads were needed on each pass.

To speed up the process, Jacquard designed his loom to run off a series of punched steel cards. The patterns of holes punched on each card told the loom which threads had to be used with each pass. By changing the order of the cards, Jacquard could change the pattern on the fabric. Looms using this invention are still in use today, and jacquard fabrics—such as damask and brocade—are named after the loom's inventor.

Jacquard's idea of using punched cards to store program information was later used by Charles Babbage, an English inventor who built the first analytical engine, the forerunner of modern computers.

THE CASE FOR COMPUTERS

It's hard to imagine a world without computers. Not only do we use them to surf the Internet, play games, and e-mail friends, but computers have also given engineers the ability to take automation to a whole new level. Before the development of modern computers, most machines in factories worked on a system of fixed automation. They were designed to do a limited number of tasks in a pre-set sequence. For example, a machine that made buttons would be set up to make only one type of button at a time. If a different type of button had to be made, the machine was stopped, and workers had to reset gears, cams, or other parts of the machine to make the new style of button. Once the new settings were in place, then the machine was started again, making only the new type of button.

After electronic computers were developed, automation took a giant leap forward. Because computers are programmable, engineers could design machines to do a wider range of jobs. Starting in the early 1970s, the fields of computer-aided design and computer-aided manufacturing (known as CAD/CAM for short) revolutionized the way that products were made. CAD/CAM devices offer flexible programming, which makes it easier and faster to change the design of a product being manufactured. One of the most important devices to use this concept is a programmable milling machine.

Like many of today's modern mechanical marvels, milling machines got their start some time ago. Eli Whitney invented the first milling machine around 1818. Here, milling refers to the process of cutting a piece of metal, wood, or some other material into a specific shape. Milling often is used to make parts with complex shapes for guns, cars, or other machines. Originally, parts with complex shapes had to be made by hand, or made with a tool called a lathe.

With a lathe, the object to be shaped is clamped between two wheels and spun quickly. A cutting blade is then pressed against the object, and material is cut away until the desired shape is reached. A lathe can make cuts in only one direction. Therefore, if a part has a complicated shape, it has to be reclamped several times in different orientations to make the proper cuts. Parts made with a lathe are usually finished by hand. And because each part is made individually, no two parts are ever exactly the same.

Whitney was trying to come up with a way of mass-producing guns for the U.S. Army. The Army needed guns that had interchangeable parts so that if a piece on one gun broke, a person could replace the part rather than the entire gun. This meant that parts had to be identical. Whitney solved the problem by turning the lathe backwards. Instead of having the material rotate, he designed a cutting blade that rotated, almost like a big drill. The material to be shaped (called a blank) was then pressed against the cutting blade at different angles.

Most modern milling machines are directed by programmable computers, which use a process called numerical control to shape parts. Numerical control means that a set of numbers are coded to correspond to points on the blank being shaped. Each point is defined by a set of three coordinates. It is something like plotting a point on an x-y axis, but there is a third coordinate (z) that marks the depth. Using the numbers, the computer instructs the machine which direction to turn the object and how far to push it against the blade. Once the machine makes all the correct cuts, the part is ejected and a new blank is inserted. Because each part is made using the same code, the parts are identical and interchangeable.

THE RISE OF ROBOTS

In addition to flexible programming, computers gave automated machines the ability to make decisions. Once this happened, machines could be turned into robots. The term *robot* comes from the Czech word *robota,* which means "forced labor." It was first used in a play written in 1920 called *R. U. R. (Rossum's Universal Robots)* to describe a group of mechanical humanoid workers who had been constructed to serve their human masters. The idea of having some type of humanlike animated machine do work for us is not new. Stories of stone or clay statues that have come to life to serve their human masters can be found in a number of ancient cultures. During the Middle Ages, artists and craftsmen in Europe built mechanical statues that used gears, pulleys and levers to move and mimic the motions of animals and people. These devices became known as "automatons," but they were used more for entertainment than as a labor force.

A robot is nothing more than an automatically operated, programmable machine that works in place of a person and has the ability to make decisions. The differences between modern automated machines and robots are getting smaller. Sometimes, it's hard to tell one from the other. A robot, however, can be "taught" new tasks. In addition, the feedback systems in robots are usually much more sophisticated. They can decide the best course of action when presented with options. This allows robots to correct many problems without a human becoming involved.

Modern industrial robots don't look anything like people. The most common type of modern robot resembles a large mechanical arm with a variety of attachments on its end. These devices can lift and turn heavy objects, weld joints, drill holes,

This robotic ultrasound monitor inspecting a welded joint is an example of a modern industrial robot. This machine, used for inspecting constructions in nuclear power plants, will detect if there are weaknesses in the welded joint.

sand metal, and even paint. The big advantage in using a robot instead of a person is that robots never get tired or bored, and they can work in dangerous environments.

In 1961, George Devol developed the first industrial robot. Within a year, General Motors began using Devol's robots on its automobile assembly lines. These early robots were limited because their memories were quite small. Most of the time, the jobs they did involved only a few simple steps, such as welding the same spot on a car body. If the robot was required to do a different job, then its entire memory had to be changed.

As computers became more powerful and their memories got larger, robots became more intelligent. Today's robots can carry out multiple jobs and be trained to do new tasks. Robots aren't used only in industry. Specialized robots defuse bombs, move radioactive waste, and even take readings inside volcanoes. Robots also have been used to explore the bottom of the ocean and the surface of Mars.

These days, developments in robotics are happening so fast that it's hard to keep up. The chance of having a robot as sophisticated as C3PO from the movie *Star Wars* is still in the future, but scientists have made humanoid robots that can walk, talk, and even climb stairs. They have also designed robots that look and act like insects, dogs, and other animals. We even have robot vacuums that will clean our floors. Robots are now being created to serve as companions for elderly people and to assist physically challenged people in their daily lives.

NANOTECHNOLOGY

When it comes to the future of tools and machines, one of the most promising new ideas lies in the development of nanotechnology. Nanotechnology is the area of engineering concerned with making devices and machines that are extremely small. It involves the manipulation of atoms and molecules to make materials, devices, and systems that are less than 100 nanometers (nm) in size. One nanometer is equal to one billionth (1/1,000,000,000) of a meter. Each hair on your head is about 10,000 nm thick!

Nanotechnology is still in its infancy. In fact, you may not even find the word in your dictionary; the term didn't exist until the 1980s. So far, the main impact of nanotechnology on our lives has been in the development of tiny materials with specialized properties. These include things like the particles in suntan lotion that block the sun and fire-retardant materials in paints and glues. In the not-so-distant future, engineers hope to start developing mechanical devices that can act as microscopic motors, switches, and sensors. The biggest advantage of nanotechnology is that smaller, lighter machines will be more energy-efficient, and once the technology is worked out, they will be cheaper to produce. One dream is to make machines that are about the same size as a living cell so that medical devices can be implanted in people's bodies using something as simple as a syringe.

As you might have guessed, moving individual atoms and molecules into the proper positions requires some highly specialized tools. Scientists have used sophisticated devices, such as atomic force microscopes and scanning tunneling microscopes. The hope is that as scientists get more familiar with working on a nano-scale, new tools and techniques will be developed that will make the manufacturing process simpler. Some day in the near future, specialized lasers might make it possible for you to build with atoms as easily as you can build with building blocks today.

SUMMING UP THE SUBJECT

As we conclude our exploration into the world of tools and machines, it's important to remember the major role they play in our day-to-day lives. Without tools, even the simplest tasks would be difficult. Jobs such as cooking, cleaning, and tending a garden would take hours longer. Without machines, we would have very few of the creature comforts that we have come to enjoy. Almost every product on which we depend—from beds and blankets to cars and computers—is mass-produced by automated machines. Tools have taken the human race from nomadic scavengers to the "movers and shakers" of our world. Tools and machines put the power of change in our hands. That power isn't always used in a positive way, though. Giant digging machines can extract coal from the ground, but they also leave the land barren. Power plants burning fossil fuels provide us with the electricity we need to run our lives, but they also pollute the air and water, and change the global climate.

As with all machines, there are trade-offs that we must consider. Sometimes we are so busy trying to figure out if we can do something, we often forget to wonder if we should be doing it. In our modern world, microprocessors and robotics have made it possible to devise machines that previously had been found only in science fiction. New discoveries in nanotechnology will almost certainly open up new ways of manufacturing that haven't even been dreamed of yet. As tools and machines continue to change, we also must look at the consequences of these new technologies.

Our world is faced with many challenges. We have to figure out how to provide clean water and food for millions of people. We have to figure out how to provide the energy to power our lives without creating a global climate catastrophe. Through the use of tools and machines, we can meet these and other challenges to make our world a better, more productive planet for all.

Glossary

axle the device which a wheel is connected to that allows it to rotate

circuit a closed loop of conductors through which electricity can flow

compound machine a device that is composed of more than one simple machine used together

conductor a substance or material that allows electricity to freely flow through it

electromagnet a temporary magnet formed when an electrical current flows through a conductor

feedback the process where information is fed back into a system to correct or change one or more operating conditions

friction the force of resistance that acts between objects when they move past each other

gauge a device used to measure a specific condition such as temperature or pressure

gear a wheel and axle in which the wheel has teeth that mesh with other gears to change the direction and speed of motion

hafting fitting a handle onto the head of a tool like an axe or hammer

inclined plane a simple machine that has a flat, tilted surface

insulator a material that blocks or slows the flow of electricity

kinetic energy mechanical energy in motion

lever a simple machine that uses a rigid bar that pivots on a fulcrum

mechanical advantage the reduction in force that is provided by a machine

meter a device used to measure the output of a machine

momentum the product of the mass of an object and its velocity

potential energy energy that is stored and is waiting to be used

programmable able to follow a series of steps or instructions to complete a task

pulley a simple machine based on the wheel and axle in which a rope, chain, or cable is carried in a groove in the wheel

resistor a device that controls (reduces) the flow of electricity in a circuit

screw a simple machine that is an inclined plane wrapped around a cylinder

sensor a device that senses some condition such as temperature or light

simple machine one of six basic devices—including the wedge, lever, pulley, screw, wheel and axle, and inclined plane—that reduces the amount of force needed to do work

spring a device that stores potential energy when it is bent or stretched

switch a device that starts and stops the flow of electricity in a circuit

tool a device used to get a specific task done.

torque force that causes an object to rotate about an axis

turbine a rotating shaft with blades on it that is turned by the pressure of a gas or liquid hitting the blades.

valve a device that controls the flow of a fluid such as air, water or steam

wedge a simple machine that is wide at one end and tapers to a point or edge at the other end

work using a force to move an object over a distance

Bibliography

Aylett, Ruth. *Robots, Bringing Intelligent Machines to Life.* Hauppauge, NY: Barron's, 2002.

Bunch, Bryan. *The History of Science and Technology.* Boston: Houghton Mifflin, 2004.

Editors of Consumers Guide. *How Things Work.* Lincolnwood, Illinois: Publications International Ltd., 1994.

Gies, Frances and Joseph Gies. *Cathedral, Forge and Waterwheel, Technology and Invention in the Middle Ages.* New York: HarperCollins, 1994.

Hewitt, Paul. *Touch This! Conceptual Physics for Everyone.* New York: Addison Wesley, 2002.

Hodges, Henry. *Technology in the Ancient World.* New York: Barnes & Noble, 1970.

Levine, Shar and Leslie Johnstone. *Mighty Machines.* New York: Sterling Publishing, 2004.

Macaulay, David. *The Way Things Work.* Boston: Houghton Mifflin, 1988.

Munro, Andrea. *The Science of Tools.* Milwaukee: Gareth Stevens Publishing, 2001.

Petroski, Henry. *The Evolution of Useful Things.* New York: Vintage Books, 1992.

Tomecek, Stephen M. *Electromagnetism and How It Works.* New York: Chelsea House, 2007.

Tomecek, Stephen M. *What a Great Idea! Inventions that Changed the World.* New York: Scholastic, 2003.

Wise, Edwin. *Robotics Demystified.* New York: McGraw-Hill, 2005.

Further Resources

Aylett, Ruth. *Robots, Bringing Intelligent Machines To Life.* Hauppauge, N.Y.: Barron's, 2002.

Bunch, Bryan. *The History of Science and Technology.* Boston: Houghton Mifflin, 2004.

Hewitt, Paul. *Touch This! Conceptual Physics for Everyone.* New York: Addison Wesley, 2002.

Levine, Shar and Leslie Johnstone,. *Mighty Machines.* New York: Sterling Publishing, 2004.

Macaulay, David. *The Way Things Work.* Boston: Houghton Mifflin, 1988.

Munro, Andrea. *The Science of Tools.* Milwaukee: Gareth Stevens Publishing, 2001.

Tomecek, Stephen. *Electromagnetism and How It Works.* New York: Chelsea House, 2007.

Tomecek, Stephen. *What a Great Idea! Inventions that Changed the World.* New York: Scholastic, 2003.

Wise, Edwin. *Robotics Demystified.* New York: McGraw Hill, 2005.

Web Sites

Antique Farming

http://www.antiquefarming.com/index.html

Explore some of the unusual tools and machines that people have used over the years, from hay rakes to steam-driven tractors.

The History of Hardware Tools

http://inventors.about.com/library/inventors/bltools.htm

Discover the history behind some of the tools that we use every day, including hammers, tape measures, and chainsaws.

Nanotechnology

http://www.ipt.arc.nasa.gov/nanotechnology.html

This site provides solid background on the science of nanotechnology and how it is being used in the space program.

The Tech Museum: Robotics

http://www.thetech.org/exhibits/online/robotics/

This site provides an online exhibition from the Tech Museum of Science, featuring the history and use of robots.

Water History.org

http://www.waterhistory.org/

Discover the history and use of waterpower over the years, from the first water wheels used to drive mills to modern turbines that generate electricity.

Picture Credits

Page

13: The Bridgeman Art Library
14: © Infobase Publishing
17: Shutterstock
18: © Infobase Publishing
24: © Infobase Publishing
30: © Infobase Publishing
33: Library of Congress Prints and Photographs Division
34: Shutterstock
36: © Infobase Publishing
37: © Infobase Publishing
42: © Infobase Publishing
47: Shutterstock
49: © Infobase Publishing
53: Dirk Ingo Franke
56: © Infobase Publishing
60: © Infobase Publishing
65: © Infobase Publishing
71: © Infobase Publishing
73: © Infobase Publishing
74: © Infobase Publishing
77: © Infobase Publishing
84: © Infobase Publishing
85: © Infobase Publishing
88: Shutterstock
91: © Infobase Publishing
92: © Infobase Publishing
96: © Infobase Publishing
103: © Infobase Publishing
109: © Hemis/Alamy
111: © Infobase Publishing
116: Shutterstock
119: © Infobase Publishing
123: © Infobase Publishing
128: © Infobase Publishing
132: Science Source/Photo Researchers, Inc.
134: © Infobase Publishing
142: © Infobase Publishing
145: © Infobase Publishing
151: © Infobase Publishing
156: © Infobase Publishing
162: Hana Kalvachova/isifa/Getty Images
164: © Washington Imaging/Alamy
168: Philippe Psaila/Photo Researchers, Inc.

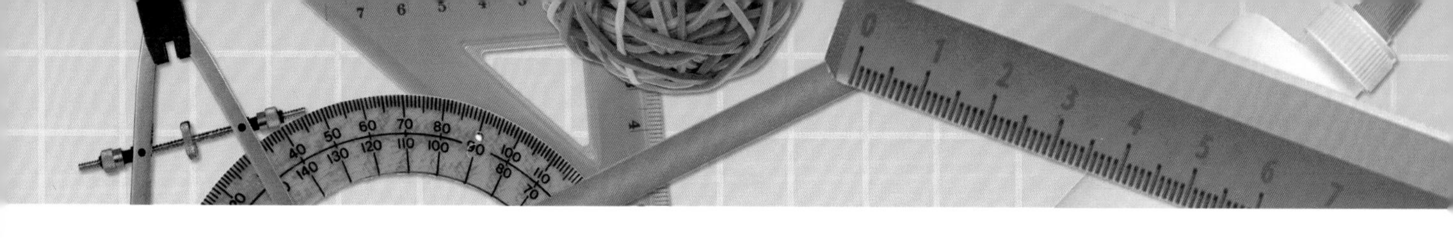

Index

About the Author

STEPHEN M. TOMECEK is a scientist and collector of old-time tools. He is the author of more than 30 nonfiction books for children and teachers, including *Bouncing & Bending Light,* the 1996 winner of the American Institute of Physics Science Writing Award. Tomecek also works as a consultant and writer for The National Geographic Society and Scholastic Inc.